EMF Effects from Power Sources and Electrosmog

Electromagnetic Frequency Sensitivities Series

Electromagnetic frequencies are everywhere in our daily lives. This brand-new series on EMF describes how having an understanding of the vast combinations of electrical and chemical problems will help in the diagnosis and treatment of electromagnetic sensitivities. This series covers the work of a renowned scientist in the field whose interests range from reversing the dysfunction of chronic illness to optimizing the health of his patients.

Air Pollution and the Electromagnetic Phenomena as Incitants is the first in an extensive series written by William J. Rea on electromagnetic sensitivity and the impact of electromagnetic phenomena on our lives. The complete list of books soon to be part of the series is as follows:

Air Pollution and the Electromagnetic Phenomena as Incitants

EMF Effects from Power Sources and Fixed Specialized Equipment: Electrosmog

The Physiological Basis of Homeostasis for EMF Sensitivity: Molds, Foods, and Chemicals

Pollutant Entry and the Body's Homeostatic Response to and Fate of the Noxious Stimuli (Dirty Electricity) from Communication Equipment

Basic Science and Science of Clinical Electromagnetic Sensitivity

Physiology of Electrohypersensitivity: EMF Treatment

EMF Effects from Power Sources and Electrosmog

William J. Rea

CRC Press
Taylor & Francis Group
Boca Raton London New York

CRC Press is an imprint of the
Taylor & Francis Group, an **informa** business

CRC Press
Taylor & Francis Group
6000 Broken Sound Parkway NW, Suite 300
Boca Raton, FL 33487-2742

First issued in paperback 2022

© 2019 by Taylor & Francis Group, LLC
CRC Press is an imprint of Taylor & Francis Group, an Informa business

No claim to original U.S. Government works

ISBN-13: 978-0-367-03109-1 (hbk)
ISBN-13: 978-1-03-233874-3 (pbk)
DOI: 10.1201/9780429020520

Library of Congress Cataloging-in-Publication Data

Names: Rea, William J. (Physician), author.
Title: EMF effects from power sources and electrosmog / by William J. Rea.
Description: Boca Raton, Florida : CRC Press, [2019] | Series:
Electromagnetic frequency sensitivities series | Includes bibliographical
references and index.
Identifiers: LCCN 2018047175| ISBN 9780367031091 (hardback : alk. paper) |
ISBN 9780429020520 (ebook)
Subjects: LCSH: Electromagnetic fields--Health aspects. | Electromagnetic
waves--Health aspects. | Electric lines--Health aspects. | Stray
currents--Physiological effect. | Electromagnetism--Physiological effect.
Classification: LCC RA569.3 R43 2019 | DDC 612/.01442--dc23
LC record available at https://lccn.loc.gov/2018047175

Visit the Taylor & Francis Web site at
http://www.taylorandfrancis.com

and the CRC Press Web site at
http://www.crcpress.com

Contents

Acknowledgments

THIS BOOK IS DEDICATED to Robert Becker, MD, an orthopedic surgeon who was one of the first clinicians to recognize the significance of EMF in medicine and surgery, and to his assistant, Andrew Marino, PhD, who helped work out the basic science of orthopedic electromagnet healing.

Recognition goes the cardiac physiologists and surgeons who worked out the basic tenets of cardiac physiology and who have the greatest advances in cardiac surgery.

This book is further dedicated to the great physiologists like Hartmut Heine, PhD, who discovered the new field of electromagnetic sensitivity—the new epidemic of the twenty-first century. This epidemic follows the problem of chemical sensitivity, for which there are 80,000 chemicals with combinations too numerous to count.

The combinations of electrical and chemical problems are so multitudinous that they are staggering to the clinician. However, a cursory understanding of the combinations of these fields can help clinicians partially understand the problems and thus help in the diagnosis and treatment of electromagnetic sensitivities.

Eventually, new treatments can be developed, eliminating illness, as shown for the healing of bones. A basic understanding of environmentally induced illness and healing must be reached by clinicians before fixed named diseases occur.

We dedicate this book to Professor Cyril Smith, PhD, at the University of Salford in England as well—along with

Dr. Jean Monro, MD, a clinician par excellence, also founder and director of the Breakspear Environmental Hospital in England. She is not only Dr. Rea's colleague and dear friend but is an innovator in keeping environmental medicine alive on the continent of Europe.

Randolph T. and Dickey L. defined the total body pollutant with chemical sensitivity that laid the foundation for understanding increased deficiency and electrical sensitivity for developing their less polluted environmental unit.

Finally, to Sam Milham, who pointed out the effect of dirty electricity on schoolchildren and teachers.

Special Recognition

SPECIAL RECOGNITION GOES TO Stephanie McCarter, M.D. and Elizabeth Seymour, M.D., who helped with reading and revising the material in this book, and to Alexis Plowden who did references, and to Gladys Morris, who did multiple typing and helped with the organization of the book as well as with many of the references.

Author

William J. Rea is a thoracic, cardiovascular, and general surgeon with an added interest in the environmental aspects of health and disease. He received his MD from The Ohio State University and was chief of thoracic surgery at the Veteran's Hospital in Dallas. He has lectured on cardiovascular and thoracic surgery at the University of Texas S.W. Medical School and on environmental sciences at the University of Texas. Dr. Rea served on the board of the American Academy of Environmental Medicine and is currently a board member of the American Board of Environmental Medicine. Dr. Rea is a fellow of the American Academy of Environmental Medicine, the American College of Surgeons, and the Royal Society of Medicine. He was part of the team who resuscitated Governor John Connally when President Kennedy was shot.

CHAPTER 1

Electromagnetic Emanations from Power Sources and Fixed Specialized Equipment

High-Tension Lines, Electromagnetic Field Towers, and Telephone Towers

E LECTROMAGNETIC EMANATIONS FROM POWER supplies and their specialized equipment are globally wide at times, often exceeding the total body pollutant load and therefore resulting in physiologic dysfunction. Each power station in the United States is connected to every other. If one area is down, electric power can be shunted from other power stations throughout the United States. This situation gives continuity of electricity over the whole country, which has been great because there have been few blackouts in the United States for years. However, this also means that people living or working near generator stations and substations, high-tension power lines, and telephone towers will be exposed to larger quantities of electric and magnetic forces because they come from all over the United States. This constant exposure to dirty electricity results in an increase in total body pollutant load and physiologic dysfunction. Not only are there more electrical generating stations and substations involved, there are thousands of permanent high-power lines and telephone facilities that put out high levels of induced electrical impulses and thus dirty electricity. This situation can be bad for people living and working near them. For example, Milham and Stetzer[1] have shown that structures and areas near cell towers have high levels of measurable dirty electricity in their electrical outlets and air.[1]

GENERAL OVERVIEW OF ELECTRICAL ENVIRONMENTAL INCITANTS (GENERATORS)

Electromagnetic field (EMF) incitants can come from various modalities and are often coupled by the coherence phenomena to the total body pollutant load of incitants such as pollen, foods, molds, mycotoxins, chemicals (natural gas, pesticides), bacteria, and viruses. Thus, the total body pollutant load can at times be so great that it makes it difficult to handle the EMF load. These EMF incitants can come from electrogenerating devices that have large capacities to generate electricity.

All transmitters, computers, compact fluorescent lights, direct current (DC) chargers, and variable speed motors contain switching power supplies.[1] *Dirty electricity is generated by arcing, sparking, and any device that interrupts the current flow.* Each interruption of current flow results in a voltage spike described by the equation $V = L \times di/dt$, where V is the voltage, L is the inductance of the electrical wiring circuit, and di/dt is the rate of change of the interrupted current. The voltage spike decays in an oscillatory manner. *The oscillation frequency is the resonant frequency of the electrical circuit, and this can be damaging to the sensitive human.*

MEASUREMENT OF DIRTY ELECTRICITY

The Graham/Stetzer (G/S) microsurge meter measures the average magnitude of the higher-frequency transients. The measurements of dV/dT read by the microsurge meter are defined as G/S units. They are a function of voltage and frequency. Thomas Edison began electrifying New York City in 1880, but by 1920, only 34.7% of all U.S. dwelling units and 1.6% of farms had electric service (Table 1.1). By 1945, 78% of all dwelling units and 32% of farms had electric service.[2] This means that in 1940, about three-quarters of the U.S. population lived in electrified residences and one-quarter did not. By 1950, the U.S. vital registration system was essentially complete, in that all 48 contiguous United States were included. Most large U.S. cities were electrified by the turn of the century,

TABLE 1.1   Growth of Residential Electric Service, U.S. 1920–1956, Percent of Dwelling Units with Electric Service

Single	All Areas in U.S.		
	City Dwellings	Farm Dwellings[a]	Urban and Rural Nonfarm Dwellings
1880	Edison discovers electricity		
1920	34.7	1.6	47.4
1925	53.2	3.9	69.4
1930	68.2	10.4	84.8
1935	68.0	12.6	83.9
1940	78.7	32.6	90.8
1945	85.0	48.0	93.0
1950	94.0	77.7	96.6
1956	98.8	95.9	99.2

Source: Modified from Milham, S. 2010. *Med. Hypotheses.* 74:337–345.
[a] 10 years longer lifetime than city dwellers in spite of more smoking in rural population.

and by 1940, over 90% of all residences in the northeastern states and California were electrified.[2]

In 1940, almost all urban residents in the United States were exposed to electromagnetic fields in their residences and at work, while rural residents were exposed to varying levels of EMFs, depending on the progress of rural electrification in their states. In 1940, only 28% of residences in Mississippi were electrified, and five other southern states had less than 50% of residences electrified. Eleven states, mostly in the northeast, had residential electrification rates above 90%. In the highly electrified northeastern states and in California, urban and rural residents could have similar levels of EMF exposure, while in states with low levels of residential electrification, there were potentially great differences in EMF exposure between urban and rural residents. It took the first half of the twentieth century for these differences to disappear.[3]

These fixed generators of electricity are permanently large stationary bases using carbon brushes to generate 60 volts of electricity and digital phone energy that are connected by *high-tension wires* or other generators, machines, or electric motors used as communication devices throughout the United States that emanate strong high-frequency dirty electricity in addition to regular electricity. These units are also bases for and/or telephones and telegraphs. Mobile phone towers can be based in fixed areas to allow transmission throughout the country. Computers, cell phones, compact fluorescent lights, halogen lamps, other electrical components, refrigerators, televisions, wireless routers, and dimmer switches can beam or plug into the air by Wi-Fi apparatus throughout an area, creating dirty electricity. Smart meters are being attached to most homes and public buildings, which can compound the problem. All of these substances can create aberrant electrical and magnetic radio frequency impulses that can alter body physiology by increasing the specific and total body pollutant load. Some will create no problems at the low frequency of 50/60 hertz. However, many of these have bursts of high-frequency dirty electricity and power surges that can and do cause aberrations of the body's physiology. These dirty electricity aberrants are usually in the gigahertz range 1–10. This aberrant impulse can cause all kinds of physiologic changes found in insects, animals, and humans.[3] These alterations in physiology may at first be imperceptible to the individual or result in subtle symptoms that are usually ignored by the individual.

When the electromagnetic load is too burdensome, the patient may develop *hypersensitivity* to electricity, which may be the harbinger of a fixed named disease such as depression, brain alterations (memory loss), insomnia, suicidal inclinations, arteriosclerosis, malignancy, neurovascular degenerative disease, vasculitis, infertility, and so on.

Overexposure to environmental incitants, both chemical and electromagnetic loads, is coupled with nutritional deficiencies and genetic metabolic defects to create these diseases and malfunctions of the twenty-first century. This combination keeps the individual from obtaining the optimum function that is desired through a lifetime.

DISEASES OF CIVILIZATION AND ELECTROMAGNETIC FIELDS

The diseases of civilization, or lifestyle diseases, include cardiovascular disease, arteriosclerosis, vasculitis, cancer, and diabetes and are thought to be caused by changes in diet, exercise habits, and lifestyle that occur as countries industrialize. *Milham[3] thinks the critical variable that causes radical changes in mortality and morbidity accompanying industrialization is electrification.* We think it is a combination of both chemicals and EMF. It is often observed that natural gas and pesticides are precursors to electrical hypersensitivity. Beginning in 1979 with the work of Wertheimer and Leeper,[4] there has been increasing evidence that some facet of electromagnetic field exposure is associated epidemiologically with an increased incidence of leukemia; certain other cancers; and noncancers like Alzheimer's disease, Parkinson's, multiple sclerosis, amyotrophic lateral sclerosis, and suicide.[4] These diseases are observed to be partially due to disruption to the cell membrane, resulting in the change of calcium channels through the membrane, allowing toxins to enter, causing disruption in normal physiology. With the exception of a small part of the electromagnetic spectrum from infrared through visible light, ultraviolet light, and cosmic rays, the rest of the spectrum is humanmade and foreign to human evolutionary experience. Milham[3] suggests that from the time Thomas Edison started his direct current electrical distribution system in the 1880s in New York City until now, when most of the world is electrified, electricity has carried high-frequency voltage transients that caused and continue to cause what are considered the normal diseases of civilization. Even today, many of these diseases are absent or have very low incidences in places without electricity.

Death rates due to tuberculosis, typhoid fever, diphtheria, dysentery, influenza, pneumonia, and measles fell sharply in this period and account for most of the decline in the causes in death rate.[3] Milham[3] shows all malignant neoplasms (Figure 1.1),

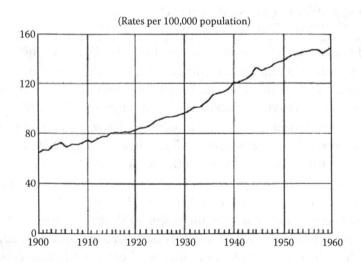

FIGURE 1.1 Death rates for malignant neoplasms: death registration states, 1900–1932, and United States, 1933–1960.

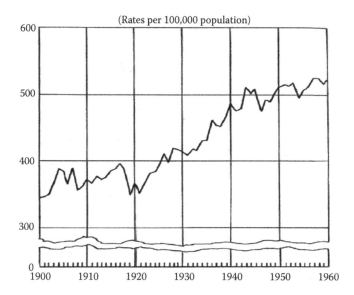

FIGURE 1.2   Death rates for major cardiovascular renal diseases: death registration states, 1900–1932, and United States, 1933–1960.

cardiovascular diseases (Figure 1.2), and diabetes (Figure 1.3) had gradually increasing death rates.[3] In 1900, heart disease and cancer were fourth and eighth in a list of the 10 leading causes of death. By 1940, heart disease had risen to first and cancer to second place, and they have maintained that position ever since. This shows that for all major causes of death examined, except motor vehicle accidents, there was a sizable urban excess in 1940 deaths. The authors of the extensive 69-page introduction to the 1930 mortality statistics volume noted that cancer rates for cities were 58.2% higher than those for rural areas.[3] They speculated that some of this excess might have been due to rural residents dying in urban hospitals. In 1940, deaths by place of residence and occurrence were presented in separate volumes. In 1940, only 2.1% of all deaths occurred for residents of one state dying in another state. The 1940 volume presents correlation coefficients for the relationship between death rates by urban or rural area of each state and the percent of residences in each state's electric service.

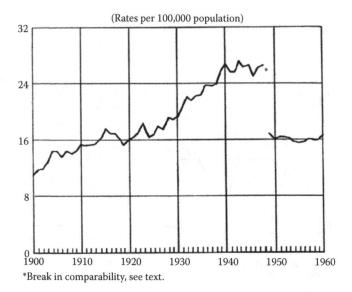

FIGURE 1.3 Death rates for diabetes mellitus: death registration states, 1900–1932, and United States, 1933–1960.

There was no correlation between residential electrification and total death rate for urban areas, but there was a significant correlation for rural areas ($r = 0.659, p = <09.0001$).

When Edison and Tesla opened the Pandora's box of electrification in the 1880s, the U.S. vital registration system was primitive at best and infectious disease death rates were falling rapidly. *City residents had higher mortality rates and shorter life expectancy than rural residents.*[5] Rural white males in 1900 had an expectation of life at birth of over 10 years longer than urban residents.[5]

Although the authors of the 1930 U.S. vital statistics report noted a 58.2% cancer mortality excess in urban areas, it raised no red flags. The census bureau residential electrification data was obviously not linked to the mortality data. Epidemiologists in that era were still concerned with communicable diseases.

Court Brown and Doll report[6] the appearance of the childhood leukemia age peak in 1961, 40 years after the U.S. vital statistics mortality data on which it was based was available. Milham[3] reports a cluster of childhood leukemia[7] a decade after it occurred, only because he looked for it. Real-time or periodic analysis of national or regional vital statistics data is still only rarely done in the United States.

The real surprise in this data set is that cardiovascular disease, diabetes, and suicide, as well as cancer, seem to be strongly related to the level of residential electrification. A community-based epidemiologic study of urban rural differences in coronary heart disease and its risk factors was carried out in the mid-1980s in New Delhi, India, and in a rural area 50 km away.[8] The prevalence of coronary heart disease was three times higher in the urban residents, despite the fact that the rural residents smoked more and had higher total caloric and saturated fat intakes. Most cardiovascular disease risk factors were two to three times more common in the urban residents. Rural electrification projects are still being carried out in parts of the rural area that was studied.[8]

It seems unbelievable that mortality differences of this magnitude could go unexplained for over 70 years after they were first reported and 40 years after they were noticed. Milham[3] thinks that in the early part of the twentieth century, nobody was looking for answers. By the time EMF epidemiology got started in 1979, the entire population was exposed to EMFs. Cohort studies were therefore using EMF-exposed population statistics to compute expected values, and case-control studies were comparing more exposed cases to less exposed controls; that is, the mortality from lung cancer in two-pack-a-day smokers is over 20 times that of nonsmokers but only 3 times that of one-pack-a-day smokers. After 1956, the EMF equivalent of a nonsmoker ceased to exist in the United States. An exception to this is the Amish, who live without electricity. Like rural U.S. residents in the 1940s, Amish males in the 1970s had very low cancer and cardiovascular disease mortality rates.[9]

If the hypothesis and findings outlined here are even partially true, the explosive recent increase in radio frequency radiation and high-frequency voltage transient sources, especially in urban areas from cell phones and towers, terrestrial antennas, Wi-Fi and WiMAX systems, smart meters, broadband internet over power lines, and personal electronic equipment, suggests that, like the twentieth-century EMF epidemic, we may already have a twenty-first century epidemic of morbidity and mortality underway caused by electromagnetic fields. The good news is that many of these diseases may be preventable by environmental manipulation, if society chooses.

A 4.5-million-dollar Air Force–supported study of pulsed 2450 MHz microwave radiation exposure of germ-free rats shows midlife immune system changes and an increase in benign and malignant tumors in the exposed rats.[10] In cows, the persistent, intermittent electrical shocks associated with stray voltage produce a typical stress syndrome characterized by increase of blood adrenal hormones and cortisol.[11–13] A recent study in mice shows that exposure for 1 hour a day for 14 days to extremely low-frequency magnetic fields (ELF-MFs) caused hyperactivity lasting for 3 months and activation of the dopaminergic D1 receptor in the brain for 1 year.[13] ELF-MF exposure measured for 1 day during pregnancy predicts asthma incidence in offspring up to 16 years later.[14] Early life stress, particularly childhood maltreatment, predicts systemic inflammation and levels of proinflammatory cytokines like interleukin in adulthood.

Evidence that neurotransmitter abnormalities are associated with disease is that there are a number of conditions for which drugs targeting neurotransmitters are used. These include but are not limited to depression, attention deficit hyperactivity disorder (ADHD), schizophrenia, Parkinson's disease, restless leg syndrome, eating disorders, anxiety disorders, insomnia, and chronic fatigue syndrome. *Electrical hypersensitivity has now been described and is rapidly increasing.* This hypersensitivity is described later in this section and certainly involves neurotransmitters as a result of the electrical and chemical situation.[15] The nonaware electrically vulnerable should take a lesson from the electrically hypersensitive in their awareness of the damaging effects of the different electrical generators of dirty electricity.

A group from Nippon Medical School in Tokyo recently reported that forest environments as compared with city environments reduce blood pressure, urinary adrenaline, noradrenaline, and dopamine and increase natural killer cell activity and expression of anticancer proteins. They thought that these effects might be due to the presence of phytoncides like α- and β-pinene in forest air.[16–18] Their findings may be due to low levels of dirty electricity in the forest as compared with city environments. These results are evidence that the *neuroendocrine and immune systems are linked and function in parallel.* Certainly, we know the case where patient reactions to these terpenes are extremely negative. The type of forest may also be highly significant when talking about the pine family or the maple or oak family of trees.

The most telling observation that electricity can be a problem involves people who live without electricity. The Old Order Amish (OOA) in North American live without electricity. They have less than half the cancer incidence of the U.S. population[19] and about half the type 2 diabetes prevalence as other U.S. citizens despite having the same body mass index.[20] Cardiovascular disease,[21] Alzheimer's disease,[22] and suicide[23] are reported to be less common in the OOA. A pediatric group practices in Jasper, Indiana, that cares for 800 Amish families has not diagnosed a single child with ADHD, and childhood obesity is almost unseen in this population.[24] Remarkably, the life expectancy of OOA has been about 72 years for the past 300 years for both men and women. In 1900, the life expectancy of U.S. males was 46.3 years and 48.3 years for females.[25] Of course, it has changed over the last half-century. If the rest of the U.S. population had the disease incidence and prevalence of the OOA, the U.S. medical care and pharmaceutical industries would collapse.

A joint Korean-US team review of 13 past studies[26] supported two previous reviews, with all three indicating a 20%–235% increase in tumors after 10 or more years of cell phone use.[26]

Over the past few years, investigators have examined cancer clusters on Cape Cod, which has a huge U.S. Air Force radar called PAVE PAWS, and Nantucket, home to a powerful LoranC antenna. Counties in both areas have the highest incidences of all cancers in the entire state of Massachusetts.[27]

The Rajasthan government has banned the installation of some mobile masts in the state. There are seven telecom operators with more than 15,000 masts in the state.[28] The new policy would bar installation of any such mast in a medical or educational institution, and permission would be given only for open spaces like parks and agricultural land.

There has to be some check on these towers, especially in residential and institutional areas. Each will be discussed separately.

RADIO FREQUENCY FROM BASE STATIONS AND TRANSMISSION TOWER SIGNS AND SENSITIVITY

Base stations and transmission towers are prime generators of EMF and radio frequency (RF) and the sources of dirty electricity. Excessive RF exposure can cause *acute problems* (headaches, insomnia, fatigue, vertigo, tinnitus, etc., and other hypersensitivity symptoms of electrohypersensitivity [EHS]). Excessive RF exposure can also cause *chronic problems* (oxidative stress and inflammation, eventually resulting in male infertility or end-stage diseases such as cancer, arteriosclerosis, and neurodegeneration). *Constant RF transmission is frequently harmful, even at low levels, and should be avoided. Frequent and repetitive intermittent transmissions are also potentially harmful, and should be avoided.* Nocturnal exposures are more problematic than daytime exposures because of RF's potential to suppress nocturnal melatonin secretion and disturb sleep, and because night is the time when individuals rest and heal from stresses (including oxidative stress). Occasional and infrequent daytime exposures are much less likely to cause an increase in chronic problems for the population at large and frequent expanding of EMF-generating apparatuses such as cell phone, computers, as is so on. Occasional and infrequent daytime exposures are still likely to provoke acute symptoms in a small percentage of the population.

RECENT DATA ON DIRTY ELECTRICITY

EMF waves can come from a variety of instruments. The evaluation of these emanating from instruments developed after the discovery of radar, but some existed before, like EMF-generating stations, motors, high-tension wires, and transformers, as previously shown. An example of dirty electricity generated from one of these new types of instruments such as TVs, cell towers, cell phones, hair dryers, and so on causing autism and a variety of diseases has been found.

STRESS SYSTEM DEREGULATION

Another example of dirty electricity effects was in chronic neurotransmitter changes in residents near a new cell tower erected in Rimbach, Austria.[29] Microwave radiation from the tower was presumed to be the active agent of the dirty electricity. Catecholamine neurotransmitters were studied in volunteers over a period of a year and a half. Epinephrine, norepinephrine, dopamine, and phenylethylamine (PEA) all had significant changes in levels, indicating chronic deregulation of the stress system, including the autonomic nervous system, from the dirty electricity. Dopamine levels dropped significantly during the first year of study. PEA levels were unchanged for 6 months and then dropped significantly over the next year. The authors postulated that cell tower radiation generated dirty electricity that resulted in a chronic stress response in the residents, accounting for the great variety of morbidity and mortality that has been reported in residents near cell towers that put out dirty electricity.[29]

RESPONSES OF THE BODY TO DIRTY ELECTRICITY

The authors of this book have found a deleterious influence in the total body pollutant load of pesticides, natural gas, formaldehyde, solvents, and many other environmental toxins, all of which cause multisystem damage, making the body's response vulnerable to stray dirty electricity. The ground regulation, immune, and neurovascular systems are the primary targets that respond through electrically charged biological substances such as neurotransmitters, hormones, and peptides, which transmit faster than the speed of light.

It has been known since the 1990s that environmental illness from biological toxins and EHS starts before and shares the process of a disastrous reduction in the nervous system enzyme cholinesterase. Its sudden depletion can cause depression and also suicidal behavior;[30] cardiac arrhythmia; and gastrointestinal, genitourinary, and respiratory dysfunction when one is exposed to dirty electricity.

All these biological processes have their inherent timing, repair, and defensive responses, especially during pregnancy and brain development, but for any biological entity process at any time. Hence, the industry "safety" standards with which government regulators collaborate are totally theoretical. The subsequently constructed safety standards for normal and dirty electricity date to 1962 when nobody knew the difference between heat-producing radiation and microwave radiation, and cell phones did not exist.

In 2002, Santini reported significant increases in such symptoms in individuals living closer than 300 meters to cell towers;[31,32] others find safety at 500 meters. The Russians say 2 kilometers, and they have been studying this the longest, so American data on 300–500 meters may be inadequate for true long-term protection.

In Poland, Bortkiewicz finds a similar increase in symptoms among residents near cell towers. Symptoms showed equal association with proximity of the tower regardless of whether the subject suspected such a casual association.[33,34] It has been observed in some areas of the United States, especially in newer building areas of business, that cell towers and cell emissions for different generators like Wi-Fi and smart meters may be within a few hundred feet, resulting in emanations from different angles that would hit an individual at the same time. We don't know the consequences of these multiple

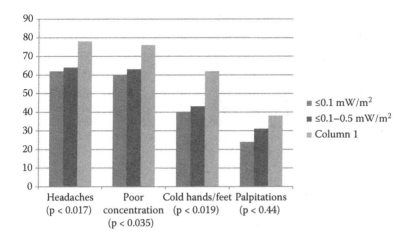

FIGURE 1.4   Percentage of subjects reporting symptoms, stratified by RF exposure levels as measured in subject's bedroom. (From Hutter, H. P. et al. 2006. *Occup. Environ. Med.* 63(5):307–313.[37])

angle bursts, but these have to be evaluated and studied carefully to be sure they don't cause brain alterations, arteriosclerosis, cancers, or cardiovascular and neurodegenerative diseases. If they trigger hypersensitivity, they will be carriers for adverse EMFs. Common sense would tell one that random frequencies from different sources might not be efficacious.

Examples of one-direction EMF from unidirectional transmissions were seen. In two studies, Abelin[35] and Altpeter[36] find evidence of disruption of sleep cycles and melatonin physiology by RF transmission during the operation and subsequent shutdown of the shortwave radio transmitter in Schwarzenburg, Switzerland.

Correlation of Signal Tower EMF Emissions and Symptoms—Austria (Figure 1.4)

In a study done in urban and rural sites in Austria, Hutter[37] finds a clearly significant correlation between exposed signal power density and headaches and concentration difficulties—despite the fact that maximum measured power densities were only 4.1 mW/m² (= 0.41 uW/cm²,) well below established "safe" limits.

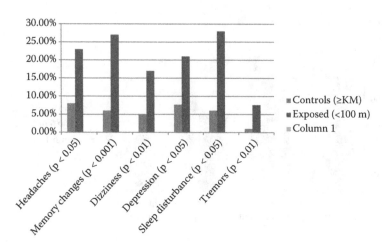

FIGURE 1.5   Percentage of subjects reporting symptoms, stratified by proximity to city's first cell phone tower. (From Abdel-Rassoul, G. et al. 2007. *Neurotoxicology* 28(2):434–440.[38])

Cell Tower EMF Emission and Symptoms—Egypt (Figure 1.5)

In Egypt, a study of inhabitants living near the first cell phone tower in the city of Shebeen El-Kom found a significant increase in headaches, memory changes, dizziness, tremors, depressive symptoms, and sleep disturbance, with lower performance on tests of attention and short-term auditory memory.[38]

Research at the military radar installation in Akrotiri, Cyprus, showed that residents of exposed villages had markedly increased incidence of migraine, headache, dizziness, and depression, and significant increases in asthma, heart problems, and other respiratory problems.[39]

Studies in Murcia, Spain, yielded similar findings, and based on measured exposures, the authors suggested that safe levels of indoor exposure should not exceed 1 uW/m² (0.0001 uW/cm²).[40,41]

In a study of residents of Selbitz, Bavaria, researchers found statistically significant increases in multiple health symptoms that demonstrated a dose-response relationship with cell phone tower transmissions. Individuals living within 400 meters of the cell phone tower had significantly more symptoms than those living >400 meters from the tower. Individuals living within 200 meters of the tower had significantly higher symptoms than those living between 200 and 400 meters from the tower.[42] Two recent reviews provide a detailed overview of research in this area.[43,44]

SOURCES OF DIRTY ELECTRICITY AND DISTANCE FROM GENERATORS

Dirty electricity readings are highest closest to the generating station and cell tower and fall away with distance. Generating stations, fixed motors, and cell towers creating dirty electricity are some of the causes of immune and nonimmune homeostatic dysfunction generated by dirty electricity. All cell towers and electrical generation stations have switching power supplies to convert the grid alternating current (AC) into direct current to operate the cell tower transmitter and charge the batteries used for backup power during grid outages. These switching power supplies interrupt the AC current flow and create dirty electricity (high-frequency voltage transients), which flows back into the grid and can cause all the entities previously discussed.[45] High-tension wires, telephone and cell towers, and substations create dirty electricity.

CELL TOWERS

An interesting example of EMF stress is shown in problems that occurred at the Olympia library close to the Rimbach cell tower in Switzerland. Here, there were only very low levels of microwave exposure in this library environment. The neurotransmitter changes in the Olympia library employees and in residents near the Rimbach cell tower were also caused primarily by cell tower dirty electricity. There was a gradual increase in urinary dopamine and PEA in the Olympia librarians. After a dirty electricity cleanup, when the cell tower was modified to decrease dirty electricity, there was a sharp contrast, seen as a decline in these neurotransmitters.[45]

The ADHD-like symptoms in children in a classroom near a cell tower were changed by modifying dirty electricity exposure, while cell tower microwave exposure was constant. Levels of a neuromodulator, β-PEA, are lower in the urine of children with ADHD.[46]

The mortality patterns linking EMF exposure to the diseases of civilization were evident long before the development of microwave transmitters in the 1940s. An Egyptian group[47] has reported that plasma adrenocorticotropic hormone and serum cortisol levels decreased over a 6-year period in people exposed to cell phones or cell phone base stations compared with controls.

Buchner and Eger's[29] surmise that the morbidity and mortality associated with cell tower EMF exposure are mediated through a chronic oxidative stress reaction seems accurate and suggests that the body recognizes EMF as a foreign invader and mounts an acute stress response to it.[29] With chronic exposure and oxidative stress, neuroendocrine and immune system deregulation results in a wide spectrum of human morbidity and mortality. Milham's[1] work shows that increasing dirty electricity in an office environment results in increased urinary levels of dopamine and PEA in exposed persons. This is evidence that dirty electricity and probably other types of EMF exposure act as initiators and propagators of illness.[29] It is clear that electromagnetic effects from power sources and specialized equipment can trigger and propagate disease processes. Changes in electrical delivery are being sought throughout the world.

MOBILE PHONE MASTS AND TELEPHONE TOWERS CAN CAUSE SEVERE PROBLEMS, AS SHOWN THROUGHOUT THE WORLD

Paralleling metallic telephone facilities, people may be exposed to high levels of induced 60 Hz. This longitudinal current is of sufficient magnitude to cause damage to equipment and personnel exposed long enough at either end because these substances also create dirty electricity. It can present a shock hazard or a chronic exposure current of this dirty electricity to personnel and people living around it. These dirty currents may cause damage to copper cable facilities as a result of overstressed dielectric levels from pair-to-pair, pair-to-shield, and shield-to-ground connections. In addition, the frequency spectrum produced by such interference will most likely disrupt communications as well as damaging the immune and nonimmune environmental receptor systems of humans. These currents reduce the level of security and reliability of circuits serving power substations and can harm individuals. The emission of stray or dirty electricity from these generating systems, substations, telephone towers, and high-tension wires can cause problems in the EMF-sensitive patient and for those who are normally functioning but susceptible. Unaware people can't perceive the devastating consequences such as cancer, arteriosclerosis, and neurodegenerative diseases that can occur long term with constant exposure to dirty electricity. High-tension stations and wires can put thousands of volts (7200) into the air, which are reduced to 220 or 120 volts for safety.

The value of reducing current provided by the power company may or may not include the effects of other supply lines, overhead ground wires, or other paralleling conductors that could reduce the effects of a power line fault but still give the adverse effects of dirty electricity. The major emphasis should be on determining the actual mutual impedance between the power distributing circuit and the telephone distributing circuits. Unfortunately, it usually isn't.

If the exposure is not reasonably parallel, the total exposure should be sectionalized and the mutual impedance of each section should be determined. The voltage sum of all sections will be the value added to the substation ground potential rise (GPR), which may exceed dirty electricity levels that are supposed to produce no health effects. Symptoms will be produced in the electrically sensitive patient, and these individuals should be listened to in order to determine the extent of the dirty electricity and its effect on humans.

Fiber optic-based cable facilities are immune to power induction and thus appear safer for the exposed individual (Table 1.2).

TABLE 1.2　Fixed Generators of High-Frequency Electromagnetic Smog (Dirty Electricity)

1. Electrogenerating stations—brushes and variable motors used as communications, split range
2. Substations
3. High-tension wires—carrying 7200 volts
4. Cell towers
5. Supply lines
6. Ground wires
7. Telephone lines
8. Mobile phone masts—wireless meters
9. Power line telecommunications
10. Radio and TV transmitters

Note: Each power station is connected to others in the United States.

Many positive changes in the environment have occurred with informed people throughout the world. However, negative changes have also occurred. For a positive example, mobile phone masts erected in County Donegal, Ireland, will have to be a least 1 kilometer away from schools (as of 2010) in order to prevent susceptibility. Also, certain frequencies have been harnessed to heal wounds and bones[48] and shock people to relieve arrhythmia and cardiac arrest. Intradermal provocation-neutralization techniques have been used on 20,000 food-, mold-, and chemically sensitive patients at the Environmental Health Center–Dallas to help heal inflammation.

Another example of positive change was when Madridiario reported that Leganes City Council, in a large town near Madrid, approved limiting mobile phone masts to 0.6 V/m or a peak power density of 0.1 microwatt per square centimeter. It also guarantees monitoring levels and providing data in real time to citizens. Surveillance will be specially controlled in sensitive places such as inside homes, workplaces, schools, hospitals, and generally any area possibly occupied by the same person for a period of more than 6 hours.[49]

In France, in May 2009, a committee (COMOP) was established to oversee experimental reduction in electromagnetism from phone masts. It was announced that 238 towns had volunteered to reduce exposure to 0.6 V/m, and 16 were chosen for the experiment. Reduction to 0.6 V/m (0.1 uW/cm²) is the preliminary stage recommended by the BioInitiative Report. The long-term aim is 0.1 V/m (0.003 uW/cm²).[50]

According to a news article in *Málaga Hoy*, there were 43 cases of cancer, 35 of which resulted in death, among the 350 inhabitants of Perez Los Cortijos living meters away from a mobile phone mast next to the watchtower in Benajarafe near the town of Velez-Málaga, Spain.[51]

In France, the director-general of Bouygues Telecom said that he was ready to set a new limit of 6 V/m indoors, as required in Italy. This would mean about 20 V/m outdoors, a considerable reduction on the obsolete hearing limits of the International Commission on Non-Ionizing Radiation Protection (ICNIRP).[51]

Limits in Salzburg, Austria, and Valencia, Spain, are 0.6 V/m outdoors (BioInitiative); in Luxemburg and some parts of Belgium 3 V/m; in Eastern Europe on average 6 V/m; in Russia 4.3 V/m; in China 6 V/m (in the process of modification); in some parts of New Zealand 1.2.75 V/m; in Switzerland 4 V/m for the 900 MHz range and 6 V/m for 1800 MHz, and for both together in multiband 5 V/m; in Lithuania, 100 times lower than in France.[51]

iBurst agreed to shut down its disputed Craigavon tower on the November 16 for 2 weeks to see if the health symptoms described by some of the residents dissipated, "including nausea, skin irritations, vomiting, headaches and sleep disorders." Those with rashes expressed that it had taken 6 weeks for the rash to heal after moving out of their homes, getting medical input, and sleeping away completely from the tower.[51]

The remaining skin texture was yet to heal even after this amount of time. Russia insists on a 2-km buffer between towers and residential properties, while New Zealand requires 500 m. Cape Town family practitioner Dr. Emdin wrote in the South African journal *Natural Medicine* in 2007: "Exposure of young children to electromagnetic field radiation may be more detrimental to their health than to adults, especially during development and maturation of the central nervous and immune systems and the critical organs." [51] Last year, Dr. Clark presented findings on a study that showed where cell phone radiation tower levels were high, 10% of that species disappeared from the landscape. [51]

HUMAN DISTURBANCE FROM GENERATORS

If there are problems in other areas where power is drawn from, like electromagnetic smog or ground currents, they may interfere with the 60 Hz generated for that structure, giving stray electricity. This connection and the generated aberrant dirty electrical activity (usually high frequency) can cause disruption of human physiology and trigger ill health such as hypersensitivity to electronic output, cancer, arteriosclerosis, and other nonmalignant degenerative diseases, especially neurodegeneration like Alzheimer's, multiple sclerosis, and Parkinson's disease. Again, these illnesses are coupled with the total and specific body pollutant load, which will measure the severity and lethality of the illness over a period of years (Figure 1.6).

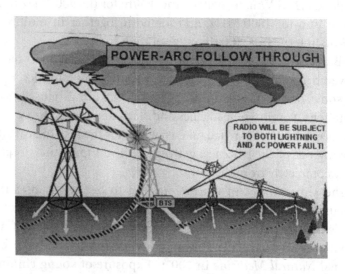

FIGURE 1.6   Dirty electricity generated while using good electricity.

Prior to 1993, laboratory studies using volunteers were confined primarily to studies of cutaneous and auditory perception and effects resulting from localized and whole-body hearing.[52] Guy et al., for example, determined a threshold for the auditory perception of pulsed RF as used in radar as 16 µJ kg^{-1} energy absorption per pulse in the head.[53] With regard to the effects of RF absorption by the whole body, this was addressed largely in the context of thermoregulation. It was known that healthy individuals can sustain an increase in body temperature up to an upper safe limit of 39°C, at which level the heart rate is considerably elevated and the sweat rate is about 1 liter per hour.[52] However, chemically and EMF-sensitive patients have difficulty in sweating and are frequently cold: this phenomenon may not apply constantly to them. In addition, early studies on the exposure of patients and volunteers to RF fields in magnetic resonance imaging systems reported that whole-body specific absorption rates (SARs) of up to 4 kg^{-1} for 20–30 minutes resulted in body temperature increases in the range of 0.1–0.5°C.[54-56]

In subsequent years, the rapid increase in wireless telecommunications, particularly those used in mobile telephony, initiated a number of research programs that included volunteer studies on the possible effects of wireless.

Milham[57] gives the history of health and disease with the onset of humanmade electrification of the earth. Though the ubiquitous distribution of toxic chemicals throughout the earth has been complete and clearly influences health both positively and negatively, as we have shown throughout these books, some of the triggering agents of illness have been found. *Chemicals have been associated with a large proportion of diseases of the twenty-first century.* Electromagnetic experience is primary in the initiation of some diseases. This EMF phenomenon, coupled with chemical overexposure, mold, mycotoxin, and microbial species invasion, is now the new complete instigator of disease (Figure 1.7).

In 2001, Ossiander and Milham[58] presented evidence that the childhood leukemia mortality peak at ages 2–4 that emerged in

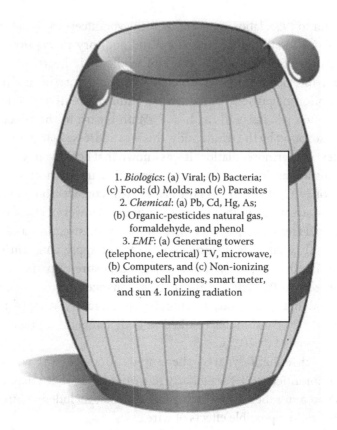

1. *Biologics*: (a) Viral; (b) Bacteria; (c) Food; (d) Molds; and (e) Parasites
2. *Chemical*: (a) Pb, Cd, Hg, As; (b) Organic-pesticides natural gas, formaldehyde, and phenol
3. *EMF*: (a) Generating towers (telephone, electrical) TV, microwave, (b) Computers, and (c) Non-ionizing radiation, cell phones, smart meter, and sun 4. Ionizing radiation

FIGURE 1.7   EMF Their Components that Increase Total Toxic (Body) Load (Burden).

the United States in the 1930s was correlated with the spread of residential electrification in the first half of the twentieth century in the United States. While doing the childhood leukemia study, Milham[58] noticed a strong positive correlation between the level of residential electrification and the death rate by state due to some adult cancers in the 1930 and 1940 vital statistics. *At the time, a plausible electrical exposure agent and a method for its delivery within residences was lacking.* However, in 2008, Milham[58] coauthored a study of a cancer cluster in school teachers at a California middle school, which indicated that high-frequency voltage transients (also known as dirty electricity) were a potent universal carcinogen with cancer risks over 10.0 and significant dose-response for a number of cancers. These carcinogens have frequencies between 2 and 100 kHz. These findings are supported by a large cancer incidence study in 200,000 California school employees that showed that the same cancers and others were in excess in California teachers statewide.

Power frequency magnetic fields (60 Hz) measured at the school were low and not related to cancer incidence, while classroom levels of high-frequency voltage transients measured at the electrical outlets in the classrooms accurately predicted a teacher's cancer risk. These fields are potentially present in all wires carrying electricity and are an important component of ground currents returning to substations, especially in rural areas. This finding of high-frequency transients helped explain the fact that professional and office workers, like the school teachers, have high cancer incidence rates. It also explained why indoor workers had higher malignant melanoma rates, why melanoma occurred on parts of the body that were never exposed to sunlight, and why melanoma rates are increasing while the amount of sunshine reaching earth is stable or decreasing due to air pollution. A number of very different types of cancer had elevated risk in the La Quinta school study, in the California school employee study, and in other teacher studies.[60] *The only other carcinogenic agent that acts like this is ionizing radiation.*[61]

OTHER GENERATORS OF TOXIC ELECTROMAGNETIC FIELDS

Among the many devices that generate dirty electricity are compact fluorescent light bulbs, halogen lamps, wireless routers, dimmer switches, and other devices using switching power supplies.[61] Any device that interrupts current flow generates dirty electricity. These will be discussed in the next section. Arcing, sparking, and bad electrical connections can also generate high-frequency voltage transients. Except for dimmer switches, most of these devices did not exist in the first half of the twentieth century. However, early electric generating equipment and electric motors used commutators, carbon brushes, and split rings, and would inject high-frequency voltage transients into the 60 Hz electricity being generated and distributed.[61]

With newly recognized electrical exposure agents and means for their delivery, Milham[58] decided to examine whether residential electrification in the United States in the first half of the last century was related to any other causes of death.[61] Most cancers showed increasing mortality in this period, and many are still increasing in incidence in the developed world.[61]

Dirty electricity can be measured using an oscilloscope of a multimeter set for peak-to-peak voltage or a microsurge meter that provides a digital readout (Graham/Stetzer units) and is easily used by nonprofessionals. G/S capacitive filters short out high frequencies and reduce transients on electrical wiring with an optimal filtering capacity between 4 and 100 kHz. (Microsurge meters and filters are available.)

Milham and Stetzer have observed that structures near cell towers have high levels of measurable dirty electricity in their electric outlets and in air.[1]

Dirty electricity was shown to be a potent universal carcinogen in a study of cancer in teachers at La Quinta Middle School in California.[59] A single year of employment at the La Quinta school increased cancer incidence by 21%. In 2010, Milham and Morgan reported a studied cluster of cancers in personnel at Vista del Monte Elementary School in North Palm Springs, CA, with a cell tower on campus within a few feet of a classroom wing. The cancer cases were overrepresented in the classroom wing closest to the cell tower. *The dirty electricity readings were highest in classrooms closest to the cell tower base and decreased linearly with distance from the cell tower base.*[62,63] Cell tower microwave radiation decreased with the square of the distance from the transmitter. A fourth-grade teacher at this school complained that her students were hyperactive and unteachable. Filtering the dirty electricity of this classroom made an immediate and dramatic improvement in student behavior. The teacher removed and plugged in the filters a number of times and reported that she could change student behavior in about 45 min.[63] At this time, the cell tower was functioning normally, and classroom microwave levels were high. *This finding suggests that the behavioral response of the students was driven by dirty electricity and not the cell tower microwaves.*

Unfortunately, historical U.S. mortality and electrification data suggest that all the so-called diseases of civilization, including cancer, cardiovascular disease, diabetes, and suicide, are caused or influenced by electromagnetic field exposure, most likely dirty electricity.[63] This was observable in U.S. mortality records very early in the twentieth century before the invention of microwaves. However, few people paid attention to it.

Hillman was involved in over 100 "stray voltage" dirty electricity cases in cows in Michigan and throughout the United States starting in about 1982.[64] Nearly all of the cases were settled out of court, thus making it impossible for the public to benefit from the findings.

RADAR

Revolutionary technology got its start during World War II. Based on primary work by British scientists, radar is credited with having allowed the British air force to survive the war. Radar was immediately put to work by air traffic controllers and the military to track moving objects and weather patterns. The first radar machines were simple affairs based on the fundamental work of British scientists who produced what were called magnetrons, or vacuum tubes. Cylinders of uniform shape, inside each magnetron sits a hollow cathode, the outer surface of which carries radionuclide metals of barium or strontium oxides in a nickel matrix. When heated through its center, this matrix emits electrons that flow through the center of the cylinder.

At a radius somewhat larger than the outer radius of the cathode is a concentric cylindrical anode. The anode serves two functions: (1) to collect electrons emitted by the cathode and (2) to store and guide microwave energy. The anode consists of a series of quarter-wavelength cavity resonators symmetrically arranged around the cathode.

A radial electric field (perpendicular to the cathode) is applied between cathode and anode. This electric field and the axial magnetic field (parallel and coaxial with the cathode) introduced by pole pieces at either end of the cathode provide the required crossed-field configuration.

Early systems depended on high-frequency radio waves to detect German planes by bouncing radio signals off objects and measuring how long it took for the signal to return, hence the name radar, for radio detection and ranging. Like many things in science, the fact that radar could be used to cook foods in what today are called microwave ovens arose from an accidental discovery.

MICROWAVE OVEN

Spencer understood that these invisible small rays could be used to move water molecules: basically to generate heat sufficient to cook anything that contained fluid. Thus was born the radar range, or microwave oven.[65]

Today, microwave ovens are standard upscale devices in kitchens around the world. Microwave ovens basically work by subjecting water or other polarized materials in food to heat generated by electromagnetic waves of about 2.4 megahertz (MHz), moving water molecules around and cooking things from the inside out. One hertz is 1 cycle per second, about the rate of the resting human heart of a well-conditioned adult. A megahertz is a million cycles per second, faster than anyone could count and the speed at which the brains of computers (microprocessors) usually work. Gigahertz is 10^9 hertz cycles per second. Modern phones work as fast as computers, sending out between 800 million and 2.4 billion cycles per second, quite close to the power needed to cook an egg.

"MICROWAVE SICKNESS"

Acute symptoms provoked by microwave radiation were first described by Russian medical researchers in the 1950s. They described a constellation of symptoms including headache, ocular dysfunction, fatigue, dizziness, sleep disorders, dermatographism, cardiovascular abnormalities, depression, irritability, and memory impairment.[66] These symptoms are not usually from foods transmitted to those of EMF sensitivity.

In the years between 1953 and 1978, the Russian government harassed the U.S. embassy in Moscow by targeting it with radiation from a microwave transmitter. Concern about health effects led to a detailed study by Lilienfeld.[67]

The abnormalities found in this study were an embarrassment to the U.S. government, since the levels of exposure experienced by embassy staff were on the order of 2 to 28 microwatts/cm^2, a level dramatically below the described U.S. safety standards for microwave exposure. The conclusions of the study were altered to soft-pedal any abnormal findings.[68,69]

However, outside epidemiologic analysis of the Lilienfeld report's published data showed that exposed embassy staff experienced a statistically significant excess of several problems, including depression, irritability, difficulty in concentrating, memory loss, ear problems, skin problems, vascular problems, and other health problems. Symptom incidence increased significantly with accrued years of exposure.[69,70] These observations were not correlated with EMF sensitivity syndrome because Smith and Monro[71] hadn't yet described it.

In 2007, Germany initiated a policy of reducing the use of computed tomography (CAT) scans because of their demonstrated cancer-causing properties and because their use increased health care costs (directly and through additional cancer incidence) by 80% in 40 years. It followed logically that last year Germany banned energy-efficient compact florescent light (CFL) bulbs because their carcinogenic radiation exceeds European exposure limits. Many jurisdictions in the United States are now framing legislation to deal both with the bulbs' unacceptable levels of radiation and the difficulty of their disposal because of the high mercury content threatening the ground water.

The European Union's September 2009 report on EMFs stressed the "serious and irreversible damage to health and environments" from EMF radiation and called on all member states to take precautionary action. Shortly afterward, the possibility of outlawing the use of cell phones for children under 18 was discussed in the EU parliament.

Switzerland, Finland, Luxemburg, and Austria supply their schools with totally EMF radiation–safe fiber optic technology for their internet and communication needs. Israel has similar legislation in the works. Those countries also set the maximum level of exposure between 5 and 10 micro W/cm^2 as "safe." [72]

Having been nearly wiped out when asbestos-related claims became undeniable in the 1990s, the cell-phone industry was informed it would not be backed when radiation hits the fan, as it must. The trigger for this decision was the United Kingdom's 2005 Stewart Commission, whose chair, citing worldwide research demonstrating harm to children's brains from cell phones, told the mobile phone industry to "refrain from promoting the use of mobile phones for children."

MICROWAVES AND MOBILE PHONES: PHYSIOLOGIC CHANGES

The intensity levels of exposure to microwaves (MWs) from mobile telephones are lower than the ICNIRP standards, which are based on thermal effects of acute MW exposures.[73] However, effects of prolonged exposure to nonthermal (NT) MWs at intensities comparable with those of mobile phones have also been observed in many studies that indicate a relationship between nonthermal microwave exposure and permeability of the brain-blood barrier,[74] cerebral blood flow,[75] stress response,[76] and neuronal damage.[77] The data obtained by the comet assay[78,79] and the micronuclei assay[80–82] imply possible genotoxic effects of NT MWs, whereas other studies did not support this genotoxicity.[83] Experimental data have indicated that nonthermal microwave effects occur depending on several physical parameters, including carrier frequency, polarization, modulation, and intermittence.[84] Differences in these physical parameters and biological variables, including genetic background and physiologic state, may explain various outcomes of studies with NT MWs.[85,86]

A recent review of available epidemiologic studies concluded that the use of mobile phones for <10 years is associated with increased risk of ipsilateral gliomas and acoustic neuromas.[87] For a long time, stem cells have been considered an important cellular target for the origination of cancer—both tumors and leukemia.[88,89] Gliomas are believed to originate from stem cells in the brain.[90] DNA double-strand breaks (DSBs) and their misrepair are critical molecular events resulting in chromosomal aberrations, which have often been associated with the origination of various leukemias and tumors, including gliomas.[91] Only one study on possible MW-induced DSBs in stem cells is available.[92] Surprisingly, the data obtained in that study by the neutral comet assay suggested that prolonged exposure time abolished the DSB formation observed at the shorter exposure time. Furthermore, the neutral comet assay has limited applicability to detect double standard breaks because similar increases in comet tails may be also caused by nongenotoxic effects that imply changes in chromatin conformation, such as relaxation of DNA loops.[93]

Digital mobile telephone radiation nowadays exerts an intense biological action able to kill cells, damage DNA, or decrease dramatically the reproductive capacity of living organisms. Phenomena like headaches, fatigue, sleep disturbances, memory loss, and so on reported as "microwave syndrome" can possibly be explained by cell death on a number of brain cells during daily exposures from mobile telephone antennas.

BREAK REPAIRS: DEOXYRIBONUCLEIC ACID

Several proteins involved in double-strand break repair, such as phosphorylated histone, 2A family member X (y-H2AX), and tumor suppressor TP53 binding protein 1 (53BP1), have been shown to produce discrete foci that colocalize to double-strand breakers of DNA referred to as DNA repair foci.[94,95] Analysis of DNA repair foci is currently accepted as the most sensitive and specific technique for measuring DSBs in untreated cells, as well as in cells exposed to cytotoxic agents.[96,97] Through analysis of the DNA repair foci in normal human fibroblasts, Markova et al.[98] were able to detect DSBs induced by a very low dose of ionizing radiation. One cGy results in only 0.4 DSB/cell on average.[98] They have also used this technique to analyze 53P1 foci in human lymphocytes exposed to MWs from global system for mobile communication (GSM)/universal global telecommunications system (UMTS) phones.[99–101]

They found that MW exposure inhibited formation of endo-genous 53BP/1-H2AX foci in lymphocytes.[99–101] This inhibition might be caused by a decrease in accessibility of DSBs to proteins because of stress-induced chromatin condensation.[100] *Inability to form DNA repair foci has been correlated to radio sensitivity, genomic instability, and other repair defects.*[102–105] Inhibition of DSB repair may lead to chromosomal aberrations by either illegitimate recombination events[102] or reduced functionality of nonhomologous end-joining.[91] Therefore, if similar effects on endogenous DNA repair foci are detected in stem cells, this might provide a direct mechanistic link to the epidemiologic data showing association of MW exposure with increased cancer risk.

STEM CELLS AND DOUBLE-STRAND BREAK REPAIR

Although y-H2AX foci have been used to analyze endogenous and induced DSBs in most studies, recent data have indicated that y-H2AX foci may also be produced by chromatin structure alternations and may not contain DSBs.[107–110] Accordingly, some y-H2AX foci may not associate with DNA damage-response proteins such as 53BP1.[98,100,101,111] High expression of endogenous y-H2AX in pluripotent mouse embryonic stem cells (−100 large y-H2AX foci per cell) was not explained by DSBs, DNA degradation, or apoptosis, but was attributed to the unusual organization of chromatin in mouse embryonic stem cells.[112] The number of endogenous 53BP1 foci (<3 foci/nucleus) appeared normal in mouse embryonic stem cells and is comparable to that found in other cell types.[112] In contrast to y-H2AX foci, which may be produced by the DSB-relevant and DSB-unrelated mechanisms, 53BP1 is relocalized to DSBs along with other DNA damage-response proteins, such as phosphorylated ataxia telangiectasia mutated (ATM), Rad50, and meiotic recombination 11 (MRE11), and there is no indication that DSB-unrelated events would result in the formation of the 53BP1 foci.[113,114] Therefore, in this study, the researchers analyzed only 53BP1 foci as a more relevant marker for DSBs.

The differences in the DSB repair pathways between mouse and human stem cells have been described.[115] In general, the comparisons of stem cells across species suggest that significant differences may be observed, so extrapolation from animal stem cell models to human health risk assessment should be done with care.[116,117] For the present study, Markova et al.[98] chose human adipose tissue–derived mesenchymal stem cells (MSCs). This cell type displays multipotency with the ability under the correct conditions to differentiate into lineages that cover a wide range of organs and tissues, such as bone, fat, cartilage, muscle, lung, skin, hepatocytes, and neurons.[118–120] Of note, MSCs are at higher risk of malignant transformation than are embryonic stem cells[89] In contrast to GSM exposure at the frequency of 915 MHz that consistently inhibited DNA repair foci in lymphocytes from 26 persons in total, GSM exposure at 905 MHz did not inhibit DNA repair focus formation, thereby providing evidence that MW effects depend on carrier frequency.[99–101] In previous studies, researchers investigated MW effects on lymphocytes. However, it would be of interest to analyze the response of human stem cells, which are usually exposed to mobile phone MWs along with differentiated human cells such as lymphocytes and fibroblasts. Therefore, in the present study, researchers exposed human stem cells and primary human fibroblasts to GSM/UMTS MWs at the same frequencies as that used previously in experiments with human lymphocytes.

Both in fibroblasts and in MSCs, γ-irradiation (3 Gy) led to significant increases in 53 BP1 foci caused by radiation-induced DSBs. In accordance with previously published data,[98] 26 foci/cell were found in fibroblasts 2 hr after irradiation, and a slightly higher level, 32 foci/cell, was detected in MSCs. These results show it is widely accepted that DNA double-strand breaks and their misrepair in stem cells are critical events in the multistage origination of various leukemias and tumors, including gliomas. Studies show whether microwaves from mobile telephones of the GSM and UMTS induce DSBs or affect DSB repair in stem cells. Markova et al. analyzed tumor suppressor TP53 binding protein 1 foci that are typically formed at the sites of DSB location (referred to as DNA repair foci) by laser confocal microscopy.

Microwaves from mobile phones inhibited formation of 53BP1 foci in human primary fibroblasts and mesenchymal stem cells. These data parallel previous findings for human lymphocytes. Importantly, the same GSM carrier frequency (915 MHz) and UMTS frequency band (947.4 MHz) were effective for all cell types. Exposure at 905 MHz did not inhibit 53BP1 foci in differentiated cells, either fibroblasts or lymphocytes, whereas some effects were seen in stem cells at 905 MHz. Contrary to this finding in fibroblasts, stem cells did not adapt to chronic exposure during 2 weeks.

The strongest microwave effects were always observed in stem cells. This result may suggest both significant misbalance in DSB repair and severe stress response. The findings that stem cells are most sensitive to microwave exposure and react to more frequencies than do differentiated cells may be important for cancer risk assessment and indicate that stem cells are the most relevant cellular model for validating safe mobile communication signals.

ELECTROMAGNETIC FIELD FILTERS

EMF Stetzerizer filters plugged into 110-volt wall outlets will dramatically reduce high-frequency currents (EMF) from the distribution circuits in the home. Also, use them in an office where computers, printers, copy machines, or other AC/DC power supplies are common. A Graham/Stetzer microsurge meter can be plugged into wall outlets to determine how much high-frequency current is in each circuit. In the Kellogg Center Ballroom, MSU and G/S readings were commonly 1600–2200. The recommended level is somewhere near 50–100 microsurge readings. The change will be obvious when you plug in the Stetzerizer filters. *The EMF filters can, however, catch fire, releasing cadmium as toxic fumes.*

There is a cluster of 35 people with childhood leukemia in a 12-mile area of Clyde, OH. This has been thought to be from dirty electricity.

Scientific evidence implies the need for reconsideration of the current exposure criteria to account for nonthermal effects, which constitute the large majority of the recorded biological and health effects. Since the mobile telephone has become part of our daily life, a better design of base station antenna networks toward the least exposure to residential areas and a very cautious use of mobile phones is necessary.

POWER LINE TELECOMMUNICATIONS

These devices produce radio interference or "noise" in the high-frequency (HF) dirty electricity range or short-wave band, just above the long-wave/medium-wave (LF/MF) bands. Especially troublesome are HomePlug Powerline Alliance (HPA) adapters, as opposed to Universal Powerline Association (UPA) ones, such as the BT Vision (Comtrend). These can cause interference at 2–30 MHz with other radio uses, such as the BBC short-wave world service, and at 13.56 MHz with RFID readers and home alarms, although some power line telecommunications (PLTs) avoid 27.12 MHz for wireless use. BT Vision Comtrend devices are said to exceed International Special Committee on Radio Interference (CISPR) regulations by 30 dB, 1000 times the maximum level to protect the radio spectrum from interference. Then how can this protect the individual? Certainly if we want to protect the electrically sensitive or vulnerable, we would eliminate such problems. It has been stated that "some of the research indicated that an access PLT system covering the whole of Greater London would significantly raise the high frequency dirty electricity noise floor in the HF bands as far away as Plymouth, while others claimed it would be detected as far away as Moscow. Of course either distance would be detrimental to some people. They also showed that near to a PLT product, HF reception could be rendered impossible for a radius of several hundred metres." This could interfere with the 20,000 amateur HF radio enthusiasts and 200,000 HF listeners in the United Kingdom, apart from MoD, security, aircraft controllers, and BBC use. "An informal analysis of reported complainants shows that victims are typically up to 150 metres from BT Vision users." [51] However, this is in a poorly informed public and medical professionals and might stretch much farther if people's perception were honed in.

PLT appears to be essentially a radio device using extended wiring antennas; it will apparently work even unplugged, and a household radio receiver will pick up HF transmissions and also VHF elements from several meters. Even BT Vision Support admitted on January 15, 2008, that "RF interference from the adapters is almost certainly what was affecting a customer's wireless keyboard. All electrical devices are subject to regulation." "It must be assumed that the mains supply already carries noise from other apparatus which may approach the limits of EN 55022, even if everything connected is in full compliance with the Directive. For PLT to operate, its signals must be greater than this minimum noise level, and so it must breach these limits, almost by definition. Yet all other mains-connected equipment, such as ITE, medical and household appliances, lighting and so forth is subject to the standard mains conducted emissions limits." [106]

In addition to acute radio interference, there is the problem of *cumulative interference*. "There is a long-distance-interference problem due to ionospheric reflection carrying PLT interference around the globe. This causes a general increase in the HF dirty electricity noise floor, to which the logical counter-response will be the environmentally undesirable use of higher radio transmitter powers." This "cumulative interference is an inevitable result of the laws of physics, and was demonstrated in practice for analogue cordless phones many years ago."[51]

Humanmade exposures, starting with the discovery of electricity to the invention of Wi-Fi and smart meters, have occurred. This area of electricity has brought great advances to civilization but also more health complications. Any of these can affect the chemically sensitive and chronic degenerative disease patient and the electrically sensitive individual. An accumulative effect may also occur, increasing the total body pollutant load and resulting in many infirmities, including nonmalignant hypersensitive chronic degenerative disease, cancer, and arteriosclerosis. Electric pollution can act jointly with other forms of pollutant—chemical, especially pesticides; natural gas; molds; mycotoxins; and foods— to cause severe reactions and disease processes.

Over the last year, there has been concern about Ofcom's failure to regulate the growing radio interference from PLT. PLT, also called broadband over power line (BPL), power line access (PLA), and power line communications (PLC), uses electronic devices that plug into mains wiring sockets to send broadband digital signals over the electrical wiring in a house. There are now said to be 0.75M such devices in use in the United Kingdom and many more in the United States. They are also known as home plug adapters.

POWER LINE COMMUNICATIONS

From a public health point of view, PLC is less problematic than RF AMI ("smart meter") communication technology. PLC could be used to reduce operating costs, train customers to conserve electricity using in-house monitors, and record and transmit time of day usage measurements to the utility.

We have turned away from the choice of PLC for two main reasons. First, it won't allow measurement of water meter readings, limiting the reduction of operating costs from elimination of meter reading. Second, PLC as currently designed does not have the bandwidth to sustain rapid "demand/response" control communications.

There are some other technical considerations that make PLC infrastructure more awkward to set up in an environment where some transmission wires are on poles and others are underground.

If "demand/response" were not on the table, and if a total bottom-line analysis of the options included the potential health costs of using RF technology, the financial analysis of the PLC option might look different than it did in the AMI business case prepared by Eugene Water and Electric Board (EWEB). A decision to read the water meters once every 3 months rather than monthly could also realize additional savings if this option were under serious consideration.

REFERENCES

1. Milham, S. and D. Stetzer. 2013. Dirty electricity, chronic stress, neurotransmitters and disease. http://apps.fcc.gov/ecfs/document/view?id=7521067073.
2. US Bureau of the Census. 1976. *The Statistical History of the United States from Colonial Times to the Present*. New York: Basic Books.
3. Milham, S. 2009. Historical evidence that electrification caused the 20th century epidemic of "diseases of civilization". http://www.stetzerizer-us.com/Historical-evidence-that-electrification-caused-the-20th-century-epidemic-of-diseases-of-civilization_df_72.html.
4. Wertheimer, N. and E. Leeper. 1979. Electrical wiring configurations and cancer. *Am. J. Epidemiol.* 109(3):273–284.
5. Vital statistics rates in the US 1940–1960, National Center for Health Statistics. Washington, DC: US Government Printing Office.
6. Court Brown, W. M. and R. Doll. 1961. Leukemia in childhood and young adult life: Trends in mortality in relation to aetiology. *BMJ*. 26:981–988.
7. Milham, S. 1963. Leukemia clusters. *Lancet*. 23(7317):1122–1123.

8. Chadna, S. L., N. Gopinath, S. Shekhawat. 1997. Urban-rural difference in the prevalence of coronary heart disease and its risk factors. *Bull. World Health Org.* 75(1):31–38.

9. Hamman, R. F., J. J. Barancik, A. M. Lilienfeld. 1981. Patterns of mortality in the Old Order Amish. *Am. J. Epidemiol.* 114(6):345–361.

10. Chou, C. K., A. W. Guy, and L. L. Kunz. 1992. Long-term, low-level microwave irradiation of rats. *Bioelectromagnetics.* 13(6):469–496.

11. Drenkard, D. V. H., R. C. Gorewit, and N. R. Scott. 1985. Milk production, health and endocrine responses of cows exposed to electrical currents during milking. *J. Dairy Sci.* 68:2694–2702.

12. Lefcourt, A. M., S. Kahl, and R. M. Akers. 1986. Correlation of indices of stress with intensity of electrical shock for cows. *J. Dairy Sci.* 69(3):833–842.

13. Shin, E-J., X-K. T. Nguyen, and T-T. L. Nguyen. 2011. Exposure to extremely low frequency magnetic fields induces Fos-related antigen-immunoreactivity via activation of dopaminergic D1 receptor. *Exp. Neurobiol.* 20(3):130–136.

14. Li, D-K., H. Chen, and R. Odouli. 2011a. Maternal exposure to magnetic fields during pregnancy in relation to the risk of asthma in offspring. *Arch. Pediatr. Adolesc. Med.* 165(10):945–950.

15. Carpenter, L. L., C. E. Gawuga, and A. R. Tyrka. 2010. Association between plasma IL-6 response to early-life adversity in healthy adults. *Neuropsychopharmacology.* 35:2617–2623.

16. Li, Q. 2010. Effect of forest bathing trips on human immune function. *Environ. Health Prev. Med.* 15(1):9–17.

17. Li Q., K. Morimoto, and H. Inagaki. 2008. Visiting a forest, but not a city, increases natural killer cell activity and expression of anticancer proteins. *Int. J. Immunopathol. Pharmacol.* 21(1):117–127.

18. Li Q., T. Otsuka, and M. Kobayshi. 2011b. Acute effects of walking in forest environments on cardiovascular and cardiovascular and metabolic parameters. *Eur. J. Appl. Physiol.* 111(11):2845–2853.

19. Westman, J. A., A. K. Ferketich, and R. M. Kauffman. 2010. Low cancer incidence rates in Ohio Amish. *Cancer Causes Control.* 21(1):69–75.

20. Hsueh, W., B. D. Mitchell, and R. Aburomia. 2000. Diabetes in Old Order Amish. Characterization and heritability analysis of the Amish Family Diabetes Study. *Diabetes Care.* 23(5):595–601.

21. Hamman, R. F., J. L. Barancik, and A. M. Lillienfeld. 1981. Patterns of mortality in the Old Order Amish. Background and major causes of death. *Am. J. Epidemiol.* 114(6):845–861.

22. Holder, J. and A. C. Warren. 1998. Prevalence of Alzheimer's disease and apolipo protein E allele frequencies in the Old Order Amish. *J. Neuropsychiatry Clin. Neurosci.* 10(1):100–102.

23. Kraybill, D. B., J. A. Hostetler, and D. G. Shaw. 1986. Suicide patterns in a religious subculture: The Old Order Amish. *J. Moral. Soc. Stud.* 1:249–262.

24. Ruff, M. E. 2005. Attention deficit disorder and stimulant use: An epidemic of modernity. *Clin Pediatr (Philadelphia).* 44:557–563.

25. http://gerontology.umaryland.edu/, Fall 2003. V6, No 2, Baltimore.

26. Myung, S. K., W. Ju, D. D. McDonnell, Y. J. Lee, G. Kazinets, C. T. Cheng, and J. M. Moskowitz. 2009. Mobile phone use and risk of tumors: A meta-analysis. *J. Clin. Oncol.* 27(33):5565–5572. doi: 10.1200/JCO.2008.21.6366.

27. Segall, M. 2011. Is Dirty Electricity Making You Sick? http://www.prevention.com/health/healthy-living/electromagnetic-fields-and-your-health

28. Rajasthan bans installation of new mobile towers. 2009. Rituraj Tiwari, ET Bureau. https://economictimes.indiatimes.com/industry/telecom/rajasthan-bans-installation-of-new-mobile-towers/articleshow/5310882.cms

29. Buchner, K. and H. Eger. 2011. Changes of clinically important neurotransmitters under the influence of modulated RF fields—a long-term study under real-life conditions. *Umwelt-Medizin-Gesellschaft.* 24(1):44–57 (Original in German).

30. Ferrie, H. The Damaging Effects of Electropollution. http://vitalitymagazine.com/article/the-damaging-effects-of-electropollution/

31. Santini, R., P. Santini, J. M. Danze, P. Le Ruz, and M. Seigne. 2002. Investigation on the health of people living near mobile telephone relay stations: Incidence according to distance and sex. *Pathol Biol (Paris).* 50(6):369–373.

32. Santini, R. S. P., P. Le Ruz, J. Danze, and M. Seigne. 2003. Survey study of people living in the vicinity of cellular phone base stations. *Electromagn. Biol. Med.* 22(1):41–49.

33. Bortkiewicz, A., M. Zmyslony, A. Szyjkowska, and E. Gadzicka. 2004. Subjective symptoms reported by people living in the vicinity of cellular phone base stations: Review. *Med. Pr.* 55(4):345–351.

34. Bortkiewicz, A., E. Gadzicka, A. Szyjkowska, P. Politański, P. Mamrot, W. Szymczak, and M. Zmyślony. 2012. Subjective complaints of people living near mobile phone base stations in Poland. *Int. J. Occup. Med. Environ. Health.* 25(1):31–40.

35. Abelin, T., E. Altpeter, and M. Roosli. 2005. Sleep disturbances in the vicinity of the short-wave broadcast transmitter Schwarzenburg. *Somnologie.* 9:203–209.
36. Altpeter, E. S., M. Roosli, M. Battaglia, D. Pfluger, C. E. Minder, and T. Abelin. 2006. Effect of short-wave (6–22 MHz) magnetic fields on sleep quality and melatonin cycle in humans: The Schwarzenburg shut-down study. *Bioelectromagnetics.* 27(2):142–150.
37. Hutter, H. P., H. Moshammer, P. Wallner, and M. Kundi. 2006. Subjective symptoms, sleeping problems, and cognitive performance in subjects living near mobile phone base stations. *Occup. Environ. Med.* 63(5):307–313.
38. Abdel-Rassoul, G., O. Abul El-Fateh, M. Abul Salem et al. 2007. Neurobehavioral effects among inhabitants around mobile phone base stations. *Neurotoxicology.* 28(2):434–440.
39. Preece, A. W., A. G. Georgiou, E. J. Dunn, and S. C. Farrow. 2007. Health response of two communities to military antennae in Cyprus. *Occup. Environ. Med.* 64(6):402–408.
40. Navarro, E., J. Segura, M. Portolés, and C. Gómez-Perretta. 2003. The microwave syndrome: A preliminary study in Spain. *Electromagn. Biol. Med.* 22(2–3):161–169.
41. Oberfeld, G., E. Navarro, M. Portoles, C. Maestu, and C. Gomez-Perretta. 2004. The Microwave Syndrome—Further Aspects of a Spanish Study. http://www.powerwatch.org.uk/pdfs/20040809_kos.pdf
42. Eger, H. and M. Jahn. 2010. Specific health symptoms and cell phone radiation in Selbitz (Bavaria, Germany)—Evidence of a dose-response relationship. *Umwelt-medizingesellschaft.* 23:1–20.
43. Khurana, V. G., L. Hardell, J. Everaert, A. Bortkiewicz, M. Carlberg, and M. Ahonen. 2010. Epidemiological evidence for a health risk from mobile phone base stations. *Int. J. Occup. Environ. Health.* 16(3):263–267.
44. Levitt, B. and H. Lai. 2010. Biological effects from exposure to electromagnetic radiation emitted by cell tower base stations and other antenna arrays. *Environ. Rev.* 18:369–395.
45. Dirty electricity, chronic stress, neurotransmitters and disease. http://apps.fcc.gov/ecfs/document/view?id=7521067073.
46. Matsuishi, T. and Y. Yamashita. 1999. Neurochemical and neurotransmitter studies in patients with learning disabilities. *No To Hattatsu.* 31(3):245–248.
47. Eskander, E. F., S. F. Estefan, and A. A. Abd-Rabou. 2011. How does long term exposure to base stations and mobile phones affect human hormone profiles? *Clin. Biochem.* 45(1–2):157–161.

48. Marino, A. A., R. O. Becker, J. M. Cullen, and M. Reichmanis. 1979. Power frequency electric fields and biological stress: A cause-and-effect relationship. *Biological Effects of Extremely Low Electromagnetic Fields.* 258–276.

49. WEEP News: Spain Will Reduce the Emissions 4,000 times/Ford, Mobile Wi-Fi Hotspots/Children's Wireless Protection Act/Effect on Electronic Equipment? 2009. http://www.madridiario.es/2009. Retrieved 2015-11-25.

50. FRANCE BioInitiative and Relay Antennas: 16 Towns Chosen for the Experimental Reduction of the Maximum EM Radiation Level to 0.6V/m. 2009. http://www.next-up.org/pdf/France_BioInitiative_Relay_Antennas_16_towns_chosen_experimental_reduction_maximum_EM_radiations_02_12_2009.pdf. Retrieved 2015-11-25.

51. Culpan a Una Antena De Telefonía Movil De 43 Casos De Cáncer. Accessed May 19, 2016. http://www.malagahoy.es/article/provincia/571081/culpan/una/antena/telefonia/movil/casos/cancer.html.

52. Electromagnetic Fields (300 Hz to 300 GHz). Environmental Health Criteria. 1993. International Programme on Chemical Safety. *WHO.* http://www.inchem.org/documents/ehc/ehc/ehc137.htm. Retrieved 2015-11-25.

53. Elder, J. A. and C. K. Chou. 2011. Auditory response to pulsed radiofrequency energy. *Bioelectromagnetics.*

54. Kido, D. K., T. W. Morris, J. I. Erickson, D. B. Plewes, and J. H. Simon. 1987. Physiologic changes during high field strength MR imaging. *Am. J. Roentgenol.* 8:1215–1218.

55. Shellock, F. G. and J. V. Crues. 1987. Temperature, heart rate, and blood pressure changes associated with clinical MR imaging at 1.5 T. *Radiology.* 163:259–262.

56. Shellock, F. G., D. J. Schaefer, and J. V. Crues. 1989. Alterations in body and skin temperatures caused by magnetic resonance imaging; is the recommended exposure for radiofrequency radiation too conservative? *Clin. Imaging.* 62: 904–909.

57. Milham, S. 2010. Historical evidence that electrification caused the 20th century epidemic of "diseases of civilization". *Med. Hypotheses.* 74:337–345.

58. Milham, S. and E. M. Ossiander. 2001. Historical evidence that residential electrification caused the emergence of the childhood leukemia peak. *Med. Hypotheses.* 56(3):290–295.

59. Milham, S. and L. L. Morgan. 2008. A new electromagnetic field exposure metric: High frequency voltage transients associated with increased cancer incidence in teachers in a California school. *Am. J. Ind. Med.* 51(8):579–586.

60. Reynolds, P., E. P. Elkin, M. E. Layefsky, and J. M. Lee. 1999. Cancer in California school employees. *Am. J. Ind. Med.* 36:271–278.

61. Milham, S. 2010. *Dirty Electricity: Electrification and the Diseases of Civilization.* 2nd ed. New York: iUniverse.

62. Milham, S. 2010a. *Dirty Electricity.* Bloomington, IN: iUniverse, pp. 78–80.

63. Milham, S. 2011. Attention deficit hyperactivity disorder and dirty electricity. *J. Dev. Behav. Pediatr.* 8:634.

64. Hillman, D., D. Stetzer, M. Graham, C. L. Goeke, K. E. Mathson, H. H. Vanhorn, and C. J. Wilcox. 2013. Relationship of electric power quality to milk production of dairy herds. *Sci. Total Environ.* 447:500–514.

65. Spencer P. 1952. Means for Treating Foodstuffs. U.S. Patent 2,605,383,605,383.

66. Liakouris, A. G. 1998. Radiofrequency (RF) sickness in the Lilienfeld study: An effect of modulated microwaves? *Arch. Environ. Health.* 53(3):236–238.

67. Lilienfeld, A. M. L. G. M., J. Cauthen, S. Tonascia, and J. Tonascia. 1979. Evaluation of health status of foreign service and other employees from selected eastern European embassies. Foreign Service Health Status Study, Final Report; Contract No. 6025-619037 (NTIS publication P8-288 163/9) pp. 1–447.

68. Goldsmith, J. R. 1995b. Where the trail leads. Ethical problems arising when the trail of professional work leads to evidence of a cover-up of serious risk and mis-representation of scientific judgment concerning human exposures to radar. *Eubios J. Asian Int. Bioeth.* 5(4):92–94.

69. Cherry, N. 2000. Evidence of Health Effects of Electromagnetic Radiation, to the Australian Senate Inquiry into Electromagnetic Radiation pp. 1–84. http://www.neilcherry.com/documents 90_m1_EMR_Australian_Senate_Evidence_8-9-2000.pdf. Retrieved 2015-11-25.

70. Goldsmith, J. R. 1995a. Epidemiologic evidence of radiofrequency radiation (microwave) effects on health in military, broadcasting, and occupational studies. *Int. J. Occup. Environ. Health.* 1(1):47–57.

71. Smith C. W. and J. Monro. 1988. Electromagnetic effects in humans. In: Fröhlich, H. (Ed.). *Biological Coherence and Response to External Stimuli*. Berlin: Springer-Verlag, pp. 205–232.

72. Ferrie, H. 2011. The damaging effects of electropollution. *Positive Health Online*. Retrieved 2015-11-25.

73. ICNIRP (International Commission on Non-Ionizing Radiation Protection). 1998. Guidelines for limiting exposure to time-varying electric, magnetic, and electromagnetic fields (up to 300 GHz). *Health Physics*. 74:494–522.

74. Nittby, H., G. Grafstrom, J. L. Eberhardt et al. 2008. Radiofrequency and extremely low-frequency electromagnetic field effects on the blood–brain barrier. *Electromagn. Biol. Med.* 27(2):103–126.

75. Huber, R., V. Treyer, J. Schuderer et al. 2005. Exposure to pulse-modulated radio frequency electromagnetic fields affects regional cerebral blood flow. *Eur. J. Neurosci.* 21(4):1000–1006.

76. Blank, M. and R. Goodman. 2004. Comment: A biological guide for electromagnetic safety: The stress response. *Bioelectromagnetics*. 25(8):642–646.

77. Salford, L. G., A. E. Brun, J. L. Eberhardt, L. Malmgren, and B. R. R. Persson. 2003. Nerve cell damage in mammalian brain after exposure to microwaves from GSM mobile phones. *Environ. Health Perspect.* 111:881–883.

78. Diem, E., C. Schwarz, F. Adlkofer, O. Jahn, and H. Rudiger. 2005. Nonthermal DNA breakage by mobile-phone radiation (1800 MHz) in human fibroblasts and in transformed GFSH-R17 rat granulosa cells *in vitro*. *Mutat. Res.* 583(2):178–183.

79. Lai, H. and N. P. Singh. 1997. Melatonin and a spin-trap compound block radiofrequency electromagnetic radiation-induced DNA strand breaks in rat brain cells. *Bioelectromagnetics*. 18(6):446–454.

80. d'Ambrosio, G., R. Massa, M. R. Scarfi, and O. Zeni. 2002. Cytogenetic damage in human lymphocytes following GMSK phase modulated microwave exposure. *Bioelectromagnetics*. 23(1):7–13.

81. Trosic, I., I. Busljeta, V. Kasuba, and R. Rozgaj. 2002. Micronucleus induction after whole-body microwave irradiation of rats. *Mutat. Res.* 521(1–2):73–79.

82. Zotti-Martelli, L., M. Peccatori, V. Maggini, M. Ballardin, and R. Barale. 2005. Individual responsiveness to induction of micronuclei in human lymphocytes after exposure *in vitro* to 1800-MHz microwave radiation. *Mutat. Res.* 582(1–2):42–52.

83. Meltz, M. L. 2003. Radiofrequency exposure and mammalian cell toxicity, genotoxicity, and transformation. *Bioelectromagnetics.* 24(suppl 6):S196–S213.
84. Belyaev, I. 2005a. Nonthermal biological effects of microwaves: Current knowledge, further perspective, and urgent needs. *Electromagn. Biol. Med.* 24(3):375–403.
85. Belyaev, I. 2005b. Non-thermal biological effects of microwaves. *Microwave Rev.* 11(2):13–29. www.mwr.medianis. net/pdf/ Vol11No2-03-IBelyaev.pdf. Accessed 2010-2-3.
86. Huss, A., M. Egger, K. Hug, K. Huwiler-Muntener, and M. Roosli. 2007. Source of funding and results of studies of health effects of mobile phone use: Systematic review of experimental studies. *Environ. Health Perspect.* 115:1–4.
87. Hardell, L., M. Carlberg, F. Soderqvist, and K. Hansson Mild. 2008. Meta-analysis of long-term mobile phone use and the association with brain tumours. *Int. J. Oncol.* 32(5):1097–1103.
88. Feinberg, A. P., R. Ohlsson, and S. Henikoff. 2006. The epigenetic progenitor origin of human cancer. *Nat. Rev. Genet.* 7(1):21–33.
89. Soltysova, A., V. Altanerova, and C. Altaner. 2005. Cancer stem cells. *Neoplasma.* 52(6):435–440.
90. Altaner, C. 2008. Glioblastoma and stem cells. *Neoplasma.* 55(5):369–374.
91. Fischer, U. and E. Meese. 2007. Glioblastoma multiforme: The role of DSB repair between genotype and phenotype. *Oncogene.* 26(56):7809–7815.
92. Nikolova, T., J. Czyz, A. Rolletschek et al. 2005. Electromagnetic fields affect transcript levels of apoptosis-related genes in embryonic stem cell-derived neural progenitor cells. *FASEB J.* 19(12):1686–1688.
93. Belyaev, I. Y., S. Eriksson, J. Nygren, J. Torudd, and M. Harms-Ringdahl. 1999. Effects of ethidium bromide on DNA loop organisation in human lymphocytes measured by anomalous viscosity time dependence and single cell gel electrophoresis. *Biochim. Biophys. Acta.* 1428(2–3):348–356.
94. Kao, G. D., W. G. McKenna, M. G. Guenther, R. J. Muschel, M. A. Lazar, and T. J. Yen. 2003. Histone deacetylase 4 interacts with 53BP1 to mediate the DNA damage response. *J. Cell. Biol.* 160(7):1017–1027.
95. Sedelnikova, O. A., E. P. Rogakou, I. G. Panyutin, and W. M. Bonner. 2002. Quantitative detection of (125)IdU-induced DNA double-strand breaks with gamma-H2AX antibody. *Radiat Res.* 158(4):486–492.

96. Bocker, W. and G. Iliakis. 2006. Computational methods for analysis of foci: Validation for radiation-induced gamma-H2AX foci in human cells. *Radiat Res.* 165(1):113–124.

97. Bonner, W. M., C. E. Redon, J. S. Dickey et al. 2008. γH2AX and cancer. *Nat. Rev.* 8(12):957–967.

98. Markovà, E., N. Schultz, and I. Y. Belyaev. 2007. Kinetics and dose-response of residual 53BP1/gamma-H2AX foci: Co-localization, relationship with DSB repair and clonogenic survival. *Int. J. Radiat. Biol.* 83(5):319–329.

99. Belyaev, I. Y., L. Hillert, M. Protopopova et al. 2005. 915 MHz microwaves and 50 Hz magnetic field affect chromatin conformation and 53BP1 foci in human lymphocytes from hypersensitive and healthy persons. *Bioelectromagnetics.* 26(3):173–184.

100. Belyaev, I. Y., E. Markovà, L. Hillert, L. O. Malmgren, and B. R. Persson. 2009. Microwaves from UMTS/GSM mobile phones induce long-lasting inhibition of 53BP1/gamma-H2AX DNA repair foci in human lymphocytes. *Bioelectromagnetics.* 30(2):129–141.

101. Markovà, E., L. Hillert, L. Malmgren, B. R. R. Persson, and I. Y. Belyaev. 2005. Microwaves from GSM mobile telephones affect 53BP1 and γ-H2AX foci in human lymphocytes from hypersensitive and healthy persons. *Environ. Health Perspect.* 113:1172–1177.

102. Bassing, C. H. and FW. Alt. 2004. H2AX may function as an anchor to hold broken chromosomal DNA ends in close proximity. *Cell. Cycle.* 3(2):149–153.

103. Celeste, A., S. Petersen, P. J. Romanienko et al. 2002. Genomic instability in mice lacking histone H2AX. *Science.* 296(5569):922–927.

104. Olive, P. L. and J. P. Banath. 2004. Phosphorylation of histone H2AX as a measure of radiosensitivity. *Int. J. Radiat. Oncol. Biol. Phys.* 58(2):331–335.

105. Taneja, N., M. Davis, J. S. Choy et al. 2004. Histone H2AX phosphorylation as a predictor of radiosensitivity and target for radiotherapy. *J. Biol. Chem.* 279(3):2273–2280.

106. Williams, T. 2007. *EMC for Product Designers.* 4th ed. Oxford: Newnes, 2007.

107. Banath, J. P., S. H. Macphail, and P. L. Olive. 2004. Radiation sensitivity, H2AX phosphorylation, and kinetics of repair of DNA strand breaks in irradiated cervical cancer cell lines. *Cancer Res.* 64(19):7144–7149.

108. Han, J., M. J. Hendzel, and J. Allalunis-Turner. 2006. Quantitative analysis reveals asynchronous and more than DSB-associated histone H2AX phosphorylation after exposure to ionizing radiation. *Radiat Res.* 165(3):283–292.

109. Suzuki, M., K. Suzuki, S. Kodama, and M. Watanabe. 2006. Phosphorylated histone H2AX foci persist on rejoined mitotic chromosomes in normal human diploid cells exposed to ionizing radiation. *Radiat Res.* 165(3):269–276.

110. Yu, T., S. H. MacPhail, J. P. Banath, D. Klokov, and P. L. Olive. 2006. Endogenous expression of phosphorylated histone H2AX in tumors in relation to DNA double-strand breaks and genomic instability. *DNA Repair (Amst).* 5(8):935–946.

111. McManus, K. J. and M. J. Hendzel. 2005. ATM-dependent DNA damage-independent mitotic phosphorylation of H2AX in normally growing mammalian cells. *Mol. Biol. Cell.* 16(10):5013–5025.

112. Banath, J. P., C. A. Banuelos, D. Klokov, S. M. MacPhail, P. M. Lansdorp, and P. L. Olive. 2009. Explanation for excessive DNA single-strand breaks and endogenous repair foci in pluripotent mouse embryonic stem cells. *Exp. Cell. Res.* 315(8):1505–1520.

113. Medvedeva, N. G., I. V. Panyutin, I. G. Panyutin, and R. D. Neumann. 2007. Phosphorylation of histone H2AX in radiation-induced micronuclei. *Radiat Res.* 168(4):493–498.

114. Yoshikawa, T., G. Kashino, K. Ono, and M. Watanabe. 2009. Phosphorylated H2AX foci in tumor cells have no correlation with their radiation sensitivities. *J. Radiat. Res (Tokyo).* 50(2):151–160.

115. Banuelos, C. A., J. P. Banath, S. H. MacPhail et al. 2008. Mouse but not human embryonic stem cells are deficient in rejoining of ionizing radiation-induced DNA double-strand breaks. *DNA Repair (Amst).* 7(9):1471–1483.

116. Brons, I. G., L. E. Smithers, M. W. Trotter et al. 2007. Derivation of pluripotent epiblast stem cells from mammalian embryos. *Nature.* 448(7150):191–195.

117. Ginis, I., Y. Luo, T. Miura et al. 2004. Differences between human and mouse embryonic stem cells. *Dev. Biol.* 269(2):360–380.

118. Bunnell, B. A., B. T. Estes, F. Guilak, and J. M. Gimble. 2008. Differentiation of adipose stem cells. *Methods Mol. Biol.* 456:155–171.

119. Porada, C. D., E. D. Zanjani, and G. Almeida-Porad. 2006. Adult mesenchymal stem cells: A pluripotent population with multiple applications. *Curr. Stem. Cell. Res. Ther.* 1(3):365–369.

120. Sasaki, M., R. Abe, Y. Fujita, S. Ando, D. Inokuma, and H. Shimizu. 2008. Mesenchymal stem cells are recruited into wounded skin and contribute to wound repair by transdifferentiation into multiple skin cell type. *J. Immunol.* 180(4):2581–2587.

CHAPTER 2

Electrosmog from Communication Equipment

INTRODUCTION

Electrosmog (dirty electricity) is a major cause of health problems in the modern world, coming not only from fixed electrogenerating equipment and specialized fixed equipment but communication equipment. The smog affects the individual by immediately resulting in symptoms (hypersensitivity) or causing delayed disease processes in the so-called nonaware vulnerable normal individual. This dirty electricity results in depression, suicidal ideations, brain dysfunction (memory loss, loss of balance, confusion episodes), arteriosclerosis, cancer, or neurovascular degenerative disease years later. It appears that the entire population is vulnerable to high-frequency generated dirty electricity emanating from communication equipment. Some people with nutrient, enzyme, immune, or genetic weakness are more vulnerable and develop maladies earlier, due to low-frequency and in-testing EMF acting as sentinels for the production of long-term chronic diseases such as cancer, arteriosclerosis, or chronic neurovascular degenerative diseases. The environmental triggering agents, other than bacteria, viruses, or parasites, are rarely looked for. Therefore, other agents such as mold, mycotoxins, foods, chemicals (pesticides, natural gas), and EMF are downplayed or ignored as the cause or agents of ill health and disease. The total body pollutant load is ignored, when in fact it is essential to understanding the etiology and developing a treatment for the environmentally wounded individual.

SOURCES OF ELECTROSMOG FROM COMMUNICATION AND MOBILE GENERATORS

According to Durham and Durham,[1] there are a variety of sources of electromagnetic transients and surges, yielding dirty electricity. Smog can interfere with the operation of electrical equipment and, in our opinion, can potentially cause damage to electrically sensitive patients[1] and unaware vulnerable individuals. These EMF-sensitive patients are canaries for the vulnerable imperceptive patient who may develop depression, suicide, mental aberrations, cancer, arteriosclerosis, and neurological-vascular degenerative disease over a period of time. See Table 2.1.

TRANSIENT ELECTRICAL GENERATORS OF DIRTY ELECTRICITY

Transient electrical output of dirty electricity can be caused by the way electrical equipment is operated, assembled, or produced, such as Wi-Fi, dimmer switches, fluorescent light, variable speed meters, cell phones, computers, smart meters, and so on. *Electromagnetic interference is generated by switched mode power supplies in computers, silicone controlled rectifier (SCR) controllers, and variable frequency drives.*[2] The interference caused by switched-mode power supplies may be corrected by increasing wire sizes, changing transformer design and configuration, and using active filters. However, many of these changes are not done as yet. Single-frequency passive filters composed of capacitors and inductors are not generally effective and shouldn't be used. The cutoff frequency

TABLE 2.1   Nonfixed Generators of Dirty Electricity and Electrosmog—Any Devices That Switch Power Supplies

1. Electrical equipment
2. Wi-Fi
3. Dimmer switches—switching from high to lower
4. Smart meters
5. Computers
6. Fluorescent lights—mainly compact bulbs
7. Switch mode position supply—dimmers
8. SRC controls
9. Cell phones
10. Variable-frequency drives
11. Transformer designs—large currents 7200 to 50–60 volts, switching of transmission lines to 50–60, 120–240 volts
12. Power surges
13. Voltage spikes or drops
14. Nuclear detonations
15. Lightning
16. Shaver chargers
17. Hair curlers
18. Microwave ovens
19. Power cable—high levels of transients
20. Plasma screens
21. Radar
22. Halogen lamps

of the filter is physically too close to the fundamental generators to allow a good filter design.[3] There are many variable transient electrical outputs, and these cause much concern and many problems for maintaining equilibrium of the electrically sensitive or vulnerable patient.

The switching of large motor loads, transmission lines, and power factor correction capacitors can create substantial noise and thus dirty electricity. For example, the reduction of large currents (i.e., from 7200 volts to 240 or 120 volts in homes or businesses) from the road power lines can cause surges on the line that damage electronics and even some machines. Electrically sensitive patients have been known to complain about these surges and conditions when they occur frequently in the average building. They appear to cause metabolic and electric disruptions in the electrically sensitive patient and apparently do so in the nonperceptive vulnerable "normal" person, who eventually develops depression, suicidal ideation, cerebral aberration, arteriosclerosis, short-term memory loss, imbalance, cancer, or neurovascular degenerative disease. However, these nonperceptive people sense and appreciate no changes until the disease is far advanced and possibly a clinical catastrophe occurs, like a severe debilitating depression, with suicide attempts, heart attack, stroke, obesity, short-term memory loss, neuromuscular atrophy, or metastasis. Some patients complain of a vibrating sensation and don't know the cause, waiting until the disease is named and perceptible, which often is too late for adequate therapy.

Sensitivity often exacerbates the problem, and the patient is unstable at this time, which exposes him or her to more intense perception of intense dirty electricity. EMF surges may be in the form of voltage spikes, voltage dips, or current transients resulting in dirty electricity. Although the energy levels for those disturbances are not as great as those from lightning, they can create problems for equipment connected to the same power lines or people close to them. Some lightning arrestors (protectors) may compensate for problems caused by switching surges on the power line and also help the electrically sensitive patient and the unaware "normal" individuals who are vulnerable to EMF changes by dirty electricity.

Radio frequency interference (RFI) can be caused by communications transmissions like power lines, computers, Wi-Fi, smart meters, fluorescent lights, cell phones, and so on. However, interference is more often a problem from the spurious emissions from electronics that are operating at a high frequency, thus generating dirty electricity. *Both digital and analog circuits can cause these emissions.* Furthermore, the RFI may be fairly broadband since it will often result from multiple subcircuits operating at the same time. Shielding is the preferred way to handle RFI. The best approach is to shield the high-frequency dirty electricity noise from escaping the source device, thus resulting in a minimum of dirty electricity and human dysfunction. Thus, the total environmental and body pollutant load can be reduced, often before low triggering of disordered metabolism.

Electromagnetic induction (EMI) is high-frequency dirty electricity noise that is picked up from a stray source and converted to a voltage on the electrical system. The sources are often the same as those generating RFI, including the aforementioned electric generators and substation power lines, computers, Wi-Fi, smart meters, fluorescent lights, cell phones, dimmer switches, variable speed meters, and so on. However, appropriate isolated grounding is an additional tool used to handle this dirty EMF noise (if the ground is not too full of EMF).

Electromagnetic pulse (EMP) interference is also derived from nuclear detonation. This is a radiated broadband electromagnetic pulse with very high energy. The energy form is very similar to lightning, but is approximately 10 times faster. Radiation is induced on all conductors and is converted into a voltage. Protectors designed for EMP will work for lightning but not necessarily for the EMF patient. However, these protectors are expensive and do not work substantially better than devices specifically designed for lightning.[4]

Lightning is a source of tremendous electric, magnetic, and thermal energy from nature, as shown previously. It is generally the predominant motivator for designing a protection system to save buildings and especially a damper for the causes of fires. This protection system is important to prevent strong dirty electricity that will not damage the human, especially the electrically sensitive canary and the electrically vulnerable.

Multiple pieces of communication equipment such as computers, fluorescent lights, dimmer switches, variable speed meters, and so on are polluters. Variable-speed motors distort 60 Hz waves, creating high-frequency dirty electricity currents.

The most common polluter is the electric power coming into the house from all the sources upstream from the utility company; mainly it is radio frequency but can also come retrograde from underground currents that have originated from dumping EMF from other homes and commercial buildings into the ground. In cities and industrial or business areas, this phenomenon has become a problem in many cases.

ATTACHMENT TO WHOLE COUNTRY

As stated previously, electrosmog gets extremely complicated not only because of this equipment and EMF flow, but because the input is attached to the whole country, and that EMF flow can be transported from another part of the country if there is a need. This situation can then create unknown electrical surges and thus high-frequency dirty electricity in addition to the equipment that it supplies.

Skin Effect

Another phenomenon observed is the *skin effect*. In the 1970s, even though we already had enough dirty electricity, the advent of computers and other electrical devices with nonlinear loads generated more dirty electricity, creating an increase in total environmental pollutant load and thus in total body pollutant load. The existing utility-neutral return wires were unable to handle the dirty electricity high frequencies returning to the substations due to the so-called skin effect. It takes a larger-diameter wire to conduct high-frequency currents, because they travel on the outside or skin of the wire. *Because of wire fires, building codes were changed to require thicker return wires in buildings, but the utilities did not change the grid neutrals.* Instead, they connected the neutral return wires to the earth by running a wire from the center tap of their transformers to the ground to use the earth as a primary neutral return to the substations.[5] This procedure led the total ground load of electricity to be markedly increased.

According to Segell,[7] the following report about electrosmog is illustrative of some of the problems brought about by the EMF of cell phones and Wi-Fi.

In 1990, the city of La Quinta, CA, proudly opened the doors of its sparkling new middle school.[6] One teacher developed vague symptoms—weakness, dizziness—and didn't return after the Christmas break A couple of years later, another developed cancer and died; the teacher who took over his classroom was later diagnosed with throat cancer. More instructors continued to fall ill, and then, in 2003, on her 50th birthday, another teacher received her own bad news: breast cancer. Milham found high levels of dirty electricity in that schoolroom.[5]

According to Milham,[5] his work has led him, along with an increasingly alarmed army of international scientists, to a controversial conclusion: the "electrosmog" that first began developing with the rollout of the electrical grid a century ago and now envelops every inhabitant of Earth is responsible for many of the diseases that impair us, as shown previously.[6]

Electromagnetic Field Behavior of Radio Frequency Smog

In the dipole effect, exposure to high-frequency electromagnetic fields from dirty electricity, with their resultant biological effects creating EMF smog, gives health consequences (100 kHz–300 GHz). Under the influence of radio frequency electric fields, electrical charges tend to accumulate on opposite cell surfaces to form induced dipoles, whose orientation changes with oscillations of the field, creating EMF smog. A dipole–dipole attraction occurs in the normal process. The attractive forces between the dipoles are enhanced when the cells are in close proximity to each other. The dipoles then align in the direction of the applied electric field and form chains of many cells or molecules. These chains are mostly single stranded, but they can be multistranded as well.

Research commenced at a time when energy-efficient devices—the major generators of dirty electricity transients—were beginning to saturate North American homes and clutter up power lines. A telltale sign of an energy-efficient device is the ballast, or transformer, that one sees near the end of a power cord on a laptop computer, printer, or cell phone charger (although not all devices have them). When plugged in, it's warm to the touch, an indication that it's damping down current and throwing off transient pollution. Two of the worst creators of transient radiation are light dimmer switches and compact fluorescent light bulbs.[7] Transients are created when current is repeatedly interrupted. A CFL, for instance, saves energy by turning itself on and off repeatedly, as many as 100,000 times per second.[7]

The human body responds to this pulsing radiation, showing that opposite charges attract, and like charges repel. When a transient is going positive, the negatively charged electrons in the body move toward that positive charge. When the transient flips to negative, the body's electrons are pushed back. These positive–negative shifts occur many thousands of times per second, so the electrons in the body are oscillating to that tune. The body becomes charged up because the individual is basically coupled to the transient's electric field.

This situation of rapid transients not only makes the body prone to cancer but can also create or exacerbate electrical sensitivity.[7] In possibly uncontrolled chemical sensitivity, studies treating the uncontrolled chemically and electrically sensitive patient are a failure at all other modalities. Physicians have found that these patients, for whom one can't get precise intradermal endpoints for antigen therapy and where the patient is reacting even to a .05-cc dose of saline, can often be stabilized by decreasing the total body pollutant load, including electrical exposure, and then these patients will also respond. They then cannot be neutralized for histamine, serotonin, capsaicin, dopamine, epinephrine, norepenephrine, foods, molds, biological inhalants, and a myriad of chemicals that have been hypersensitive responses to aberrant dirty electricity and have lost their specific endpoint and normal EMF projections. They can also be neutralized for EMF effects by various specific EMF frequencies generated by a frequency generator. With the reduction of transient oscillations, the tissue stabilizes and the dilution endpoints hold for months, allowing proper desensitization and normalizing the body's metabolic response.

Some people want to remove the health dangers of Wi-Fi, WiMAX, digital TV, and digital radio. When shielded, the patient loses sensitivity and eventually becomes well. Therefore, the departments of the Drome and Ardeche in France aim to replace wireless with fiber-optic cables that stabilize these frequencies and oscillations. This will cover 100% of the population of 95 million, connecting 372,000 homes via ordinary phone lines through 213 switchboards.[8]

It will provide an ultra-high-speed broadband connection (100 Mbps) without any loss of signal due to distance, for a "triple play" service of internet, telephone, and TV. Wi-Fi and WiMAX will stop altogether because of their health dangers.[8]

Another example of wiping out dirty electricity, the demolition of Entel PCS's mobile phone mast in the O'Higgins district of Santa Cruz, was confirmed in Chile. This was on the grounds that the structure "violates the constitutional rights" of people affected by its radiation.[8] This idea followed the Institute of Public Health's report acknowledging that human health is damaged by the harmful effects of the radiation produced by cellular phone masts.

Charlie, a Sydney brain cancer surgeon, urges people to put mobile phones on loudspeaker and wait until microwaves have finished beeping before opening them.[8] He recommends that "all electrical appliances be put at the foot of the bed and not the head of the bed"[8] due to his observations on the adverse effects of dirty electricity.

Havas[9] showed from a double-blind study of 25 volunteers that some people react to DECT phone exposure with an increased heart rate.

France and other countries are joining Eastern European countries in beginning to accept the increasingly established scientific evidence of the biological dangers of EMF. Even telecom companies are now volunteering to move toward biological limits as the evidence becomes increasingly incontrovertible and is being accepted legally and by insurance companies.

As with cordless phones and Wi-Fi, unless the user deliberately switches the units off, in which case the products are compliant, they will be putting out the full signal level that can be harmful to the body 24 hours a day. Some units, however, are said to be capable of sending intermittent check signals in standby mode.[8]

Compact Fluorescent Lights and Radio Frequency

CFLs and other low-energy lights can emit RF radiation. An ordinary household radio receiver can, for instance, pick up RF interference at 150 kHz at a distance of over 1 meter from a particular CFL. Some GE electronically ballasted CFLs carry a warning against use "near maritime safety equipment or other critical navigation or communication equipment operating between 0.45 and 30 MHz."[8] However, no warning comes for the individual.

"Artificial lights" are limited to specific analyses of the ultraviolet radiation subtypes UVA, UVB, and UVC. But many electrical sensitive (ES) sufferers feel adverse effects near CFLs even when visible light from a CFL is shielded, suggesting the cause may not be UV but RF radiation from the CFL.[8]

Some common devices emitting RF pollution detected by a household radio receiver, in addition to external signals from radio and TV transmitters, include dLAN Homeplugs and associated wiring, compact fluorescent lights, fluorescent strip lights at the end, microwave ovens, TV monitors (a cathode ray tube [CRT] also has a magnetic field), garage remote-controlled door apparatuses, chargers for the Phillips battery shaver, and hairdryers. In contrast with the shaver charger, a Nokia mobile phone charger produced very high magnetic fields but little RF pollution detected by a household radio receiver.[8]

Illness from radio exposure has been known since the 1930s, and increased cancer levels have been known since Milham's[10] study of amateur radio operators (often using HF radio) in 1988 and Hocking's 1996 study[11] of TV masts. Some microwave pulses have RF elements, such as GSM, UMTS (3G), and Bluetooth, while WiMAX is mainly RF.[8] Dirty electricity, where transients in the kHz and MHz range ride on household electric wiring, is becoming very common, from PLTs, plasma screens, compact fluorescent lights, TVs, microwave ovens and light dimmers. Comtrend warns against using PLT Homeplug adapters with microwave ovens and hairdryers, both known to produce RF interference. The health dangers associated with high levels of radio frequency range [RFR] transients on power cables is increasingly well known following the 2008 study by Milham and Morgan.[12]

In the 1980s, CRT visual display units (VDUs) (TV monitors), which produce RF pollution, were often associated with skin dermatitis and other skin problems. Compact fluorescent lights can also cause skin problems. In 2002, links were first made between radio transmitters and melanoma. Huttanen's study of 2009 showed through spontaneous muscular movement that the human body is particularly sensitive to radio frequencies, as in radio and TV transmitters, probably because of wavelength and resonances.[13]

Homeplug dLAN devices are good insofar as they remove the need for dangerous Wi-Fi. They are probably better than Wi-Fi regarding long-term health, since Wi-Fi may radiate further and at higher levels.[8]

Coghill[8] worries about increasing electrosmog in the United Kingdom. *The amount of radiation in the atmosphere has increased massively in the last 20 years.* It affects DNA, and cells. About 7% of the UK population has electrosensitivity. A Chinese man in a paddy field is better protected from this radiation than a child at school. In the UK more people will become electrosensitive and, ultimately, some will die of cancer.[8]

Some reports suggest that the switch-on of digital TV and enhancement of city Wi-Fi are causing problems.[8] Phillips[14] suggests it is absurdly complacent to pretend that these electromagnetic fields are not going to have any impact on health. Far from doing no harm, some studies suggest that as much as 5% of the population may already be suffering from headaches, concentration difficulties, chronic fatigue, irritability, and behavioral problems because of this electrosmog. According to the CDC the incidence of Brain cancer increased by 0%–4% each year from 2001–2014 and leukemia is the only cancer more common than brain cancer in young people.[172] In 2003 Dr. O'Neill[15] suggests that "brain tumors are on the increase, reportedly in the region of 2% per year. Brain tumor cases nearly doubled in the last year."[8] In an attempt to overcome arrhythmia, people have tried to develop shields.

The biggest problem had been arrhythmia from prolonged ELF (power lines, household current) as well as radiofrequency exposures; that [lined under several layers of cloth with laminated aluminum foil] reduced those episodes to rare occurrences, typically after spending much time in the vicinity of cell phone users. When the antennas are below these higher transmissions, the hat does not help, and a complete suit might be necessary. A secondary benefit of shielding was the near elimination of long-standing allergies, including one to cats from early childhood. The patient was sensitive to EMF all along, and no one had any idea. It took the advent of cell phones to point out the dangers of EMF, except one has to deal with Wi-Fi and/or cell phone addicts. When that is unavoidable, a silver-mesh veil is attached to the hat or draped over the body.

Mobile phones appear to be one of the generators of electromagnetic smog. They rival fixed EMF generators of electricity in their output of dirty electricity.

BASE STATION ANTENNAS FOR GENERATING SIGNALS VS. MOBILE PHONE SIGNALS

According to Panagopoulos and Margaritis,[16] EMF radiation from base station antennas is almost identical to that from mobile phones of the same system (GSM or distributed control system [DCS]), except that it is approximately several times up to several hundred times more powerful. Thus a large abundance of EMF smog occurs. Thereby, effects produced by mobile phones at certain distances can be extrapolated to represent effects from base station antennas at distances about 100 times longer. Another difference is that handset signals include one pulse per frame occupying one time slot, whereas base station signals include, again, one pulse per frame, but this pulse may occupy 1–8 time slots depending on the number of subscribers at each moment. In other words, the ratio between pulse peak power and time-averaged power is usually higher for handset mobile phone signals compared to base station signals.[16-21] These facts explain the variability of sensitivity to EMF waves and mobile phones seen in EMF-sensitive patients. Of course, if the sensitive individual is around base stations and uses a mobile phone, he or she will be getting an increased dose of radiation, which can cause physiological disruption and body malfunction.

In epidemiological studies among mobile phone users, a major difficulty is the variation of parameters governing the exposure from hand-held mobile phones, that is, the distance from the nearest base station, which can considerably change the intensity of the radiation emitted by the phone; the actual duration of daily use; and so on. Nevertheless, the studies done on habitant living close to base stations are more consistent since the station emits a more constant radiation level on a daily basis and therefore a person residing nearby receives measurable radiation at least for several hours per day.

A recent Egyptian study[22] found that inhabitants living near mobile telephone base stations may develop a number of neuropsychiatric problems like headaches, memory changes, dizziness, tremors, depression, and sleep disturbances, also reported in previous studies as a "microwave syndrome."[23] There were changes in the performance of neurobehavioral functions. Most of these sensitivities are seen in the EMF-sensitive patient. Similar results were found by other studies in different countries like France,[24] Poland,[25] Spain,[23] and Austria.[26]

It has been shown that the vast majority of health effects of digital mobile telephony radiations are nonthermal, and a lot of biological effects were recorded at radiation intensities much lower than the values of efficient safety criteria. This is why several countries in Europe have established much more stringent national exposure criteria, like Italy, Poland, Russia (10Î¼W/cm²), or Austria (0.1Î¼W/cm²; EMF worldwide standards).

Mobile Phone: Animal Studies

According to Panagopoulos and Margaritis, there are already a very large number of published clinical and epidemiological studies regarding research on possible health risks from cellular mobile telephony radiations.[16] A large and increasing number of studies (biological, clinical, and epidemiological) have recorded a variety of nonphysiological changes with increased probabilities for health hazards, including several types of cancer. Inconsistencies observed between studies are partly expected since no identical conditions can ever be attained between different studies and different labs, but also they are explained by some authors to be due to biased samples. However, there is bias in any study, so this fact will often be ruled out as a dormant sister in EMF sensitivity.

According to Panagopoulos and Margaritis,[16] microwaves are found to produce, thermally and nonthermally, a large number of biological effects in many cellular and animal studies.[27] In the case of radiation emitted by mobile telephony smog-saturated antennas at intensities to which people are normally exposed, the effects are nonthermal, as verified by different experimenters.[28-34]

Regarding nonthermal effects of RF radiation, it is a must to refer to the works of Bawin et al. and Blackman et al.[36,37] from the 1970s and 1980s, although these works relate to lower frequency RF radiation. In those experiments, RF radiation with carrier frequencies 147 and 450 M(10^6) Hz modulated by sinusoidal ELF signals at 0–40 Hz (Hz is 10^3 10 gute/sec) was found to decrease Ca^{++} concentration in chicken brain cells The effect was found to be greatest at modulation frequencies 6–20 Hz and intensities 0.6–1 mW/cm².[35,36] Nonmodulated RF signals were not found to be as bioactive as modulated ones by ELFs, and these effects were found to be nonlinearly dependent on radiation intensity and frequency, exhibiting windows within which the phenomena appeared and then disappeared for outside values.[37,38]

Long-term exposure of rats to 900 MHz mobile phone radiation produced oxidative stress (increased oxidant products of free radicals) in retinal tissue. Melatonin and caffeic acid phenethyl ester (CAPE) components of honeybee propolis administered daily to the animals prior to their electromagnetic radiation (EMR) exposure caused a significant reduction in the levels of the oxidant products.[39] In a previous study of the same group, melatonin was found to reverse oxidative tissue injury in rat kidneys after 10 days exposure at 30 min per day to 900 MHz GSM radiation emitted by mobile phones.[40]

Male mice were exposed to 1800 MHz GSM-like microwaves, 0.1 mW/cm², for 2 weeks on workdays, 2 h per day. Red blood cell count and serum testosterone levels were found to be significantly higher in the exposed groups, but no significant alterations were found in the other investigated variables.[41]

Mice prone to the development of lymphomas, exposed for two 30-min periods per day for up to 18 months to 900 MHz pulsed microwave radiation with a 217 Hz pulse repetition frequency at SAR ranging from 0.007 to 4.3 W/kg, developed twice the number of tumors than unexposed ones.[42]

Male Wistar 35-day-old rats were exposed to 2.45 GHz radiation for 2 h/day for a period of 35 days at a power density of 0.344 mW/cm² (SAR 0.11 W/kg). The study revealed a decrease in protein kinase C (PKC) activity. Electron microscopy study showed an increase in mobile telephony radiation effects in 113 of the glial cell population in the exposed group.

The results indicated that *chronic exposures may affect brain growth and development.*[43] In another study of the same group, single-strand DNA breaks were measured as the tail length of a comet. The study showed that chronic exposure to microwave radiation at nonthermal levels (SAR 1 and 2 W/kg) causes a statistically significant increase in DNA single-strand breaks in rat brain cells.[44]

In another study, mice placed within an RF smog–saturated antenna park were repeatedly mated five times while they were continuously exposed to very low levels of RF radiation (0.168–1.053 μW/cm²). A progressive decrease in the number of newborns per maternal mouse was observed after each mating, which ended in irreversible infertility.[45] In a more recent study of the same group, it was found that exposure of pregnant rats to GSM-like 940 MHz radiation at 5 μW/cm² resulted in aberrant express of bone morphogenetic proteins (BMPs)—major endocrine and autocrine morphogens known to be involved in renal development—in the kidneys of newborn rats.[46] An increase in the number of micronuclei in rat bone marrow erythrocytes, a sign of genotoxicity, was observed after 30 days exposure for 2 h daily to 910 MHz microwave radiation.[47]

In several other mammal studies, no effects were found with regards to the genotoxicity of second-generation mobile telephony and third-generation universal mobile telecommunication system radiation.[48–52]

The evidence is rather consistent and suggests that mobile phone–type RF exposure has no effect on auditory function in rodents. It is also clear that animals can hear the pulsed RF characteristic radar above given thresholds through a thermoelastic expansion mechanism. If an increase in permeability occurs in the blood–brain barrier or peripheral blood tissue barriers and an efflux of K^+ occurs, it can cause symptoms of potassium deficiency. In addition, Na^+ and Ca^{++} go into cells and release Ca^{++} stored in the cells. This Ca^{++} combines with protein kinases A and C and, when phosphorylated, increases cell membrane sensitivity 1000 times. This process or condition results in mold, food, chemical, and electrical sensitivity.[53]

Human Effects

Due to the many complaints about problems with cell phones, Wi-Fi, and smart meters, shielding has been proposed. Fiber optics has been suggested to solve the dirty electricity problem.

Chattanooga: The Smarter City

The Electric Power Board (EPB) in Chattanooga, Tennessee, is one of the first community-owned utilities to install a 100% fiber-optic network, which uses the fiber-optic network for smart grid applications in addition to the triple-play media services (i.e., high-speed internet, video, and telephone) EPB already provides (Figure 2.1).

While many utilities struggle with the question of whether to build a smart grid, for the Electric Power Board in Chattanooga, it was never an issue.

Virtually unlimited bandwidth gives EPB lightning-fast, two-way communications with every device in its distribution system. While a network this robust is overkill for metering, EPB realized that fiber is essential for tightly coordinated load-shedding activities, the split-second responsiveness required in distribution automation, and a virtual real-time energy management tool for customers.

For instance, because bandwidth is no problem, EPB is able to offer its customers simultaneous internet upload and download speeds of up to 1 gigabit—200 times faster than the current national average and 10 times faster than the Federal Communications

FIGURE 2.1 EPB technician deploys smart meter at customer home. Photos courtesy of EPB.

Commission's (FCC) National Broadband Plan (a decade ahead of schedule). This is just one example of how EPB's 100% fiber-optic network will provide countless benefits for the community in terms of education, healthcare, economic development, quality of life, and more.

The network is designed so that data move efficiently between the utility and every endpoint, regardless of whether a premise is connected directly to the fiber-optic network. This configuration ensures a smooth transition to implementing the energy efficiency initiatives proposed by the Tennessee Valley Authority (TVA), such as time-of-use pricing, load shedding, customer signaling, and advanced distribution automation applications, that are becoming more prevalent (Figure 2.2).

EPB has built fiber optics throughout their entire customer service area, and communications services are now available to all homes and businesses. By the end of 2012, all 170,000 homes and businesses were equipped with a Tantalus smart meter. Although building the network is the first step, the smart grid, not TV, internet, and phone services, drove the business case. It was clear

FIGURE 2.2 EPB operations center—built to handle the massive amounts of data generated by the smart grid.

this technology was the key to increasing reliability and managing energy costs well into the future.

Both the smart grid and communications benefits of a 100% fiber-optic infrastructure are paying off by attracting new business.

Chattanooga will also be a test bed for electric vehicles (EVs), with the Tantalus network providing the means through which 300 street-side charging stations will be monitored. During peak conditions, EPB can avoid overloading transformers by deferring charging until the evening. On the other hand, the network can be used to measure how much power is being withdrawn from EV batteries if EPB needs to access energy stored in car batteries.

While discussion about the smart grid centers on how homeowners can better manage consumption and mitigate cost, EPB sees local industries as having the most to gain. Access to reliable, low-cost power is essential.

High-bandwidth and low-cost reliable power are two things that can make a community more attractive to industries looking to relocate. "Most of these savings can be accounted for because with the upgraded switching scheme, businesses don't suffer nearly the degree of productivity loss that would otherwise result from a prolonged power outage." (Harold DePriest, EPB's President & CEO)

When time-of-use prices are in effect, manufacturers can reschedule or resequence operations to periods when energy is the cheapest. Furthermore, the ability to optimize power quality benefits the manufacturing process by minimizing the impact caused by minor fluctuations in voltage or frequency, which once disrupted highly sensitive manufacturing equipment.

Future-Friendly Network
Advantages

1. Network leverages EPB's fiber-optic investment for triple-play media and enhanced smart grid functionality

2. One of the first cities to implement municipally owned 100% fiber-optic network; economic driver for the region and opportunity to improve energy efficiency

3. Fiber-optic network supports simultaneous upload and download of up to 1 gigabit

4. Fast, low-cost deployment of 1500 smart meters per week; self-configuring, self healing network streamlines smart meter implementation and subsequent upgrades

5. Helps industry become more efficient; time of use (TOU) pricing expected to result in savings of $2.3 million a year for 22 manufacturers involved in time-of-use rate pilot

6. Anticipates 40% in outage reductions resulting from improved distribution system management and intelligent switch technology

7. Easy scalability; can support communications with the millions of data points, including meters, demand response devices (i.e., smart thermostats and load control devices), distribution equipment, and smart appliances

EPB doesn't expect to be in the pole position as the country's fastest broadband city for too much longer. Other communities across the country are in the process of building gigabit networks. But it's really a global race, with a handful of other cities around the world boasting lightning-fast broadband service (Figure 2.3).

MOBILE PHONE EFFECTS ON HUMANS

Gene Damage and Heat Shock Proteins—Mom's Cell Phone Use Retards Baby's Speech

Chronic cell phone use interrupts mother/child bonding. In the first 6 months of their lives, babies of moms who talk more on cell phones do not bond, look, or coo as much as those whose moms pay more attention. Chronic cell phone usage can mean maternal

FIGURE 2.3   Fiber pull-down in Chattanooga. EPB is building a lightning-fast, two-way fiber-optic network that provides communications with every device in its distribution system.

distraction from infant needs and tuning out cues about newborn needs.

EMF smog caused repairable DNA damage, and increased expression of heat shock protein (Hsp 70) without changes in cell proliferation rates was detected in human lens epithelial cells after 2 h exposure to a 1.8 GHz RF field amplitude modulated at 217 Hz with 3 W/kg SAR. The DNA damage was determined by use of the comet assay.[54]

Increased expression of genes encoding ribosomal proteins and consequently upregulating the cellular metabolism in human cell types was found after in vitro exposure to 900 and 1800 MHz mobile phone radiation.[55] In Panagopoulos's study, gene and protein expression were altered in human endothelial cell lines after 900 MHz GSM mobile phone radiation exposure at an average SAR of 2.8 W/kg. Genes and proteins were differently affected by the exposure in each of the cell lines, suggesting that cell response to this type of radiation might be genome- and proteome-dependent, which in turn might explain to some extent the discrepancies in replication studies between different laboratories.[56]

Exposure of human endothelial cells *in vitro* to GSM 900 MHz mobile phone radiation for 1 h at nonthermal levels, average SAR 2 W/kg, caused a transient increase in heat shock protein (hsp27) phosphorylation and transient changes in protein expression levels.

Rapid (within minutes) induction of heat shock protein (hsp70) synthesis was found in the insect *Drosophila melanogaster* after *in vivo* exposure to GSM 1900 MHz mobile phone radiation.[57]

According to a theoretical report, repetitive stress caused by mobile phone radiation leads to continuous expression of heat shock genes in exposed cells and tissues and may result in cancer induction.[58]

Two hours of exposure by cellular mobile phone changed the structural and biochemical characteristics of acetylcholinesterase, resulting in a significant alteration of its activity. The enzyme was exposed within an aqueous solution at 5 cm distance from the mobile phone.[59] Exposure of myoglobin solution to 1.95 MHz microwave radiation for 3 h at nonthermal levels was found to affect the folding of the protein and thereby change its biochemical properties.[60] Thus, properties when the folding isn't precise effect adaptation to the changing environment and in the mitochondria function, endoplasmic reticulum, and cell function, resulting in disease.

In vitro exposure of human skin fibroblasts to GSM radiation for 1 h induced alterations in cell morphology and increased heat expression of mitogenic signal transduction genes, cell growth inhibitors, and genes controlling apoptosis.[61] In an earlier study, with a 960 MHz GSM-like signal at SAR 0.021, 0.21 and 2.1 mW/cm^2 with exposure times 20, 30 and 40 min respectively, Panagopoulos et al.[24] found a decrease in the proliferation rate of transformed human epithelial amnion cells. The maximum effect was reached at lower power level with a longer exposure time than at higher power levels.[62] In another study, *in vitro* exposure of human peripheral blood lymphocytes to continuous 830 M (10[6]) Hz radiation, with average SAR 1.68.8 W/kg, was found to produce loss and gains of chromosomes (aneuploidy), a somatic mutation leading to cancer. The effect was found to be activated via a nonthermal pathway.[63] Though all of these studies were found to be abnormal, they were *in vitro*. Familiar findings were seen *in vivo* when exposed to EMF smog.

Central Nervous System Effects

Clinical studies on humans exposed to EMF smog by using mobile telephone radiation have found it frequently affects electroencephalograms (EEG), electrodermal activity (EDA), and the synthesis rate of hormones like melatonin in humans. In a series of early experiments performed by a Finnish group, GSM mobile phone smog exposure was found to alter the EEG oscillatory activity of healthy adult subjects in the 6–8 and 8–10 G(10[9])Hz frequency bands during cognitive (visual memory) tasks. In more recent experiments of the same group, exposure of 10–14 year old children to mobile phone GSM field smog while performing an auditory memory task induced changes in their brain oscillatory activity. Obviously, the effects of EMF smog can be detrimental to some groups with selective exposure. EEG responses occurred in the smog frequencies 4–8 Hz and 15 Hz.[64]

Exposure for 30 min to pulse-modulated 900 MHz mobile phone-like EMF smog increased waking regional cerebral blood flow (rCBF) and enhanced EEG power in the alpha frequency range (8–12 Hz) prior to the onset of and during sleep. Exposure to the same field without pulse modulation did not enhance power in waking or sleep EEG.[65] In another set of experiments by the same group, 30 min exposure to the same 900 MHz GSM-like field during the waking period preceding sleep increased the spectral power of the EEG in non–rapid eye movement sleep. The maximum increase occurred in the 9.75–11.25 Hz and 12.5–13.25 Hz frequency ranges during the initial part of sleep. Since exposure during waking modified the EEG during subsequent sleep, the changes in the brain function induced by mobile telephone radiation are considered to outlast the exposure period.[65]

Mobile phone exposure prior to sleep was found to decrease rapid eye movement and sleep latency and to increase EEG spectral power in the 11.5–12.5 Hz frequency during the initial part of sleep following exposure.[66] Some other studies have failed to find any effects of mobile phone-microwave exposures on EEG during cognitive testing or to replicate earlier findings.[67,68]

Mobile phone radiation was found to affect the evoked neuronal activity of the central nervous system (CNS) as represented by EDA, an index of the sympathetic nervous system. Mobile phone exposure was found to lengthen the latency of EDA (skin resistance response) irrespective of the head side next to the mobile phone.[69] Therefore, mobile phone exposure may increase the response time of users with different negative consequences, for example, an increase in the risk of phone-related driving hazards and so on. At EHC-Dallas, electrically sensitive patients have been found to have altered heart rate variability, as measured by an autonomic nervous system (ANS) apparatus, no matter the frequency.

A statistically significant increase of chromosomal damage was found in blood lymphocytes of people who used GSM 900-MHz mobile phones, compared to a control group of nonusers matched according to age, sex, health status, drinking and smoking habits, working habits, and professional careers. The increase was even greater for users who were smokers or alcoholics.[70] Those people had previously already had their pollutant load increased.

In another type of clinical study, exposures of humans to GSM 900-MHz and DCS 1800-MHz mobile phone fields for 35 min were not found to change arterial blood pressure or heart rate significantly during or after the exposure.[71] Prolonged use of mobile phones (more than 25 min per day) was found to induce a reduction in melatonin production among male users. The effect was enhanced by additional exposure to a 60-Hz ELF magnetic field.[72]

Two studies about possible immediate to short-term effects of GSM- and UTMS (third generation of mobile networks)-like exposure on well-being and cognitive performance in humans based on questionnaires found contradictory results. The first[73] reported no effects of GSM-like exposure, while UTMS-like exposure was found to reduce well-being and cognitive performance. The second[74] reported no effects at all from either type of radiation.

According to Hardell et al.,[75] the concluding results of up-to-date epidemiological studies among users for more than 10 years use of mobile phone use consistently indicate an increased risk of acoustic neuroma and glioma, especially for ipsilateral exposure. Earlier work of the same research group found a connection between digital (second generation) and analog (first generation) mobile phone use and malignant brain tumors, highest for a latency period more than 10 years.[76]

Central Nervous System: Measurable Brain Changes
Blood–Brain Barrier

The brain of mammals is protected from potentially dangerous materials by the *blood–brain barrier*, a specialized neurovascular complex. The blood–brain barrier functions as a selective hydrophobic filter that can only be easily passed through by small fat-soluble molecules. Other non–fat-soluble molecules, for example, glucose, can pass through the filter with the help of carrier proteins that have a high affinity for specific molecules.

It is known that a large number of disorders of the central nervous system are caused by disturbances of the barrier function of the blood–brain barrier,[77] such as Parkinson's disease, Lou Gehrig's disease, multiple sclerosis, and neuropathy. Clearly, the chemically sensitive are prone to intoxication, as are most electrically sensitive patients, apparently where the membrane channels or dynamic holes change.

Severe warming of the brain can lead to an increased permeability of the blood–brain barrier for those materials whose passage should actually be prevented, such as chemicals that give blood–brain barrier dysfunction. The results of first experiments with high-frequency fields of high intensity, which led to a higher permeability of the blood–brain barrier, were then interpreted as a consequence of warming by the HF radiation.

However, Table 2.2 lists a whole series of studies in which greatly increased permeability of the blood–brain barrier was produced through pulsed high-frequency fields of very low intensity,[77–80] among others with carrier and modulation frequencies that corresponded to those of mobile telephony (GSM), and no heating occurred.

Items 4 and 5 of Table 2.2 correspond to the altered route that odors follow in the chemically sensitive in contrast to normals. The chemically sensitive senses odor impulses up the olfactory nerve to the frontal lobe and then to the hippocampus. This is in contrast to normal people, where the odor impulses go directly to the hypothalamus and hippocampus. This alteration could change the choline and other chemical intake, causing chemical sensitivity. This change, along with albumin influx, may allow multiple chemical changes, allowing for EMF and chemical sensitivity changes.

TABLE 2.2   Nonthermal High-Frequency Fields That Increase the Permeability of the Blood-Brain Barrier

1. Mobile telephones—GSM
2. High frequency–low intensity chemical change
3. Continuous 2.45 GH_2 with power flux 50–100 w/m² decrease noradrenalin in hypothalamus
4. Same intensity and high frequency—odors in Pons and medulla increase dichytroxy phenylacetic acid, 5-OH oxyloacetic acid
5. 2045 GH_z field—modulated by 500 #2 pulse modulators, frontal cortex and hippocampus decrease choline uptake

Blood–Brain Barrier Change with Mobile Phones
Several studies have reported that microwave exposures increase the permeability of the blood–brain barrier (BBB).[33] A Swedish group has reported that 915 MHz microwaves at nonthermal intensities cause leakage of albumin into the brain cells through the BBB in rats, accumulating in the neurons and glial cells that surround the capillaries in the brain.[77] The same group reported that GSM mobile phone radiation from a test mobile phone with a programmable constant power output opens the BBB for albumin, resulting in damage of brain cells in rats. The power density and SAR were within the ICNIRP limits.[77] These were the first experiments that indicated cell damage caused by mobile phone radiation, although this radiation was not a real mobile phone signal. However, in an earlier study of the same group, continuous-wave and pulsed 915 MHz radiation at relatively high intensities, 1 and 2 W, respectively, was not found to damage the brain or promote brain tumor development in rats.[77] Exposure of an *in vitro* BBB model consisting of rat brain cells growing in a culture with pig blood cells, exposed to 1800 MHz microwave radiation pulsed at a 217 Hz repetition rate (DCS-like) at SAR 0.3–0.46 W/kg increased the permeability to sucrose of the BBB twice compared to the control culture. No significant temperature rise was detected during the exposures.[33] In a latter study of the same group, *in vitro* exposure of three other BBB models with distinctly higher barrier tightness than the previously used one did not cause any effect on the permeability of the BBB of the models.[82]

Deoxyribonucleic Acid Damage from Mobile Phones
In regards to DNA damage or cell death induction due to microwave exposure, in a series of early experiments, rats were exposed to pulsed and continuous-wave 2450 MHz radiation for 2 hours at an average power density of 2 mW/cm2a, and their brain cells were subsequently examined for DNA breaks by comet assay. The authors found a dose-dependent (0.6 and 1.2 W/kg wholebody SAR) increase in DNA single- and double-strand breaks

4 hours after the exposure to either the pulsed or continuous-wave radiation.[83,84] The same authors found that melatonin and Ntert-butyl-alpha-phenylnitrone (PBN), both known free radical scavengers, block the above effect of DNA damage by microwave radiation.[85] Although these experiments were the first to report DNA damage by microwaves, the radiation intensity (2 mW/cm^2) was relatively high, exceeding the international exposure limits (ICNIRP), and the radiation frequency was the same as in microwave ovens. *In vitro* exposure of mouse fibroblasts and human glioblastoma cells to 2450,[86] 835.2, and 87.74 MHz[87] radiation at SAR 0.6 W/kg was not reported to damage DNA as measured by comet assay.

A number of recent studies have reported DNA damage, cell damage, or cell death induced by mobile telephony smog or similar RF radiation at nonthermal intensity levels,[28,30,77,88–90] while some other studies did not find any such connection.[91–95] Aitken et al.[88] reported damage to the mitochondrial genome and the nuclear betaglobin locus in the spermatozoa of mice exposed to 900 MHz 0.09 W/kg SAR smog for 7 days, 12 h per day. Diem et al.[28] reported single- and double-strand DNA breakage in cultured human and rat cells exposed to 1800 MHz mobile phone–like radiation smog. Panagopoulos et al.[30] found DNA fragmentation at a very high degree caused in the reproductive cells of female *Drosophila* insects only by a few minutes of daily exposure to a real mobile phone signal for only few days. *These were the first experiments that showed extensive DNA damage and cell death by real digital mobile phone GSM and DCS signals.* Previous experiments of the same group had shown a large decrease in the reproductive capacity of the same insect caused by real mobile phone similar exposures.

Cell Phone Use and Wireless Digital Enhanced Cordless Telecommunication

Multiple studies of cell phone users in the last decade found evidence of a similar pattern of symptoms to be provoked in some users.[96–108]

Animal Studies

The animals of this group (n = 6) were exposed to radiation within their home cages 3 hours per day for 8 months. The exposure protocol of "3 h/day × 8 months" was chosen in order to mimic daily typical mobile phone operation by an active person. The mobile phone was placed underneath the cage. A semi-Faraday cage was specially constructed with one open surface to allow mobile phone communication and at the same time to prevent radiation leakage toward sham-exposed animals. The GSM 900 MHz electrical field intensity of the radiation emitted by the mobile phone was measured using the Smartfield meter (EMC Test Design, LLC, Newton, MA, USA), placing the dual-band omnidirectional probe (900, 1800 MHz) inside a similar cage housing the animals positioned at the same place either at the end or beginning of exposure. The obtained measurements were reproducible on a daily basis (minimum-maximum value depending on the sound intensity). In order to simulate the conditions of the human voice and activate mobile phone ELF-modulated EMF emission, a radio station was played as a source of auditory stimulation throughout the exposure time. The measured electrical field intensity was below ICNIRP's recommendations,[109] within the range of 15–22 V/m in the various areas within the cage, with the animals also following the typical GSM power modulation by sound intensity. The SAR value (SAR = $o^{-*}E2/p$), calculated as previously described,[110,111] was between 0.17 and 0.37 W/kg. This is a rough estimation of the whole body average SAR of individual animals.

Another example of dirty electricity occurred with mobile telephone and Wi-Fi. According to Ferrie,[112] the detrimental health effects of cell phones, Wi-Fi exposure from Wi-Fi equipped schools and TVs, electromagnetic fields, and dirty electricity from the microwave-based technology including the microwave ovens from which they receive their meals daily makes worrying about our kids experimenting with drugs, sexual escapades, alcoholic binges, or a touch of crime insignificant. In comparison to the

Wi-Fi-equipped schools they are sent to, the cell phones, the hours they spend in front of the TV (forget the content—the TV itself is worse), and the microwave oven from which they often receive their meals are also detrimental because they generate dirty electricity.[112]

World Health Organization Director General Gro Harlem Brundtland (1998–2003), a convener of the World Commission on Environment and Development in 1987, did not permit any cell phones at the WHO's Geneva headquarters because they caused her debilitating headaches, which has been seen in many ES and chemically sensitive (CS) patients. Of course, she had the wrath of the industry descend upon her accordingly. She is currently suffering from cancer. In 2004, the WHO defined electrohypersensitivity as a phenomenon where individuals experience adverse health effects while using or being in the vicinity of devices emanating dirty electric, magnetic, or electromagnetic fields, a sometimes debilitating problem occurring several orders of magnitude under the limits of internationally accepted standards.[112]

Everybody is at risk, according to Milham,[12] who first presented childhood leukemia clusters in the 1970s that eventually proved beyond any doubt that the risk for this disease is directly related to the presence of dirty electricity emanating from power lines and subsequently also cell phone towers. *Those areas of the world that have little or no electricity have almost no incidence of leukemia.* In addition to leukemia, evidence now supports that cell phones cause, especially, brain tumors; cancer of the eye, salivary glands, and testicles; and the development of autism and ADHD. While several causes are known for all of these conditions, research has confirmed that the use of cell and cordless phones and laptop computers speed up their manifestation such that people in their 30s are now beginning to be diagnosed with advanced Alzheimer's from the dirty electricity emanating from these apparatuses. The degranulation process of live brain cells when exposed to cell phone radiation has been experimentally demonstrated. Cancer incidence is also significantly higher within 400 meters of a cell phone tower or transmitter site. Havas may now have discovered a third type of diabetes caused exclusively by EMR.

There are cases where a household contained a meter and it registered electricity from a neighbor's smart meter, dramatically increasing the ground currents underneath the house.

The degranulation process of live brain cells when exposed to cell phone radiation has been experimentally demonstrated. Cancer incidence is also significantly higher within 400 meters of a cell phone tower or transmitter site.[112]

People who talk on mobile phones are up to five times more likely to develop brain tumors than those who stick to landlines.[113]

A number of previous studies into mobile phone safety "substantially underestimated"[113] the cancer risks and that tumors are much more common on the side of the head to which the mobile is held than on the other side.

After reanalyzing the earlier studies, it was concluded that the risk of these tumors is between 10% and 500% higher with long-term mobile phone use.[113] These included a Swedish study that originally concluded that people who used mobile phones for at least 10 years were 3.9 times more likely to develop an acoustic neuroma compared to those who rarely or never used the devices.[113] The reanalysis, designed to take into account flaws in the design of the study that could have skewed the results, put the increased risk at 4.9 times.[113]

Another study had concluded that 10 years of mobile phone use raised the odds of "same-side" gliomas by 60%.[113]

Reanalysis of a third study concluded that every 100 hours of mobile phone use raises the odds of meningioma by 24%.[113]

The results of some of these studies were included in the world's largest study on mobile phone safety. The most vulnerable to dirty electricity are children, pregnant women, human brains in general, testicles, and ovaries. This was confirmed in 2007 by the World Health Organization and the International Agency for Research on Cancer and in 2010 by a Swedish government study, all showing that cell phone use increases the chances of brain cancer by 40%.[113]

Last year, the *European Journal of Oncology* reported that serious heart and related problems (e.g., arrhythmia, palpitations, heart flutter, racing heartbeat, fainting, profuse sweating, etc.) can occur with pulsed radiation from cell phones and other devices as low as 0.5% of the existing Canadian and U.S. federal guidelines, which permit 10 mill micro/Wm². The truly science-based exposure guidelines demonstrate that nobody should be exposed to more than 1 microW/m².[113]

A declassified U.S. Navy document from 1971 summarizes the more than 2300 studies on health effects from microwave radiation.[113] Of course they knew—microwave technology was already part of the military arsenal intended to inflict bodily harm. The Military University in Germany developed EMF protection standards for civilian building codes. In addition to people, building materials require protection from rapid corrosion caused by radiofrequencies and microwaves. Germany's radiation protection body also advises its citizens to use landlines instead of mobile phones and warns of "electrosmog" from a wide range of other everyday products, from baby monitors to electric blanklets.[113]

Digital mobile telephony radiation nowadays exerts an intense biological action able to kill cells, damage DNA, or dramatically decrease the reproductive capacity of living organisms. Phenomena like headaches, fatigue, sleep disturbances, memory loss, and so on reported as "microwave syndrome" can possibly be explained by cell death on a number of brain cells during daily exposure from mobile telephony antennas (Table 2.3).

TABLE 2.3 Digital Mobile Phones and Microwaves Trigger

1. Changes in intracellular ionic concentration
2. Synthesis ratio of different biomolecules
3. Changes in cell proliferation rates
4. Changes in reproductive capacity
5. Changes in gene expression
6. DNA damage
7. Reduction of melatonin
8. Cell death

Physiology Related to Mobile Phones
Scientific evidence implies the need for reconsideration of the current exposure criteria to account for nonthermal effects, which constitute the large majority of the recorded biological and health effects. *Since mobile telephony has become part of our daily lives, a better design of base station antenna networks toward the least exposure to residential areas and a very cautious use of mobile phones are necessary.*[16]

Panagopoulos and Margaritis[16] have shown a number of serious nonthermal biological effects ranging from changes in cellular function like cell proliferation rate changes or gene expression changes to cell death induction. There is also a decrease in the rate of melatonin production and changes in electroencephalogram patterns in humans, population declines in birds and insects, and small but statistically significant increases of certain types of cancer. These are attributed to the radiation emitted by mobile telephony antennas of both handsets and base stations.[16]

Khurana et al.[114] found that 8 of 10 studies reported increased prevalence of adverse neurobehavioral symptoms or cancer in populations living at distances <500 meters from base stations. None of the studies reported exposure from dirty electricity above accepted international guidelines, suggesting that current guidelines may be inadequate in protecting the health of human populations.

Over the past four decades, cell phone types and uses have changed radically. The first bulky, heavy phones relied on analog signals that were basically on all the time without interruption, putting out around 600 milliwatts (mW) per second. By the late 1980s, an array of such systems ran around the world, including the Advanced Mobile Phone System (MPS) in North America, Asia Pacific, Russia, Africa, and Israel in the frequency band between 800 and 900 MHz and the Nordic Mobile Telephone (NMT) 900 system since 1986 in Scandinavia, Netherlands, Switzerland, and Asia. Radiation from these big old phones was believed to produce mostly heat.

Digital devices of newer 2G and 3G phones can do the work of a handheld personal computer and can handle and receive much more information with a fuller range of multimedia services. Launched only about two decades ago, these smaller, sleeker phones rely on low-rate digital signals powered at around 250 or 125 mW encoded at 217 pulses/sec. These 3G and 4G phones also entail smaller waves at lower frequencies. They are basically on all the time, constantly and continually signaling to base stations to get new information. As a result, despite their smaller wavelengths, with their constant signaling, 3G phones can result in greater cumulative average exposure to radiofrequency signals through the usual far-ranging multimedia use.

The main reason gorillas in Africa have been reduced in numbers by 90% over the past two decades is their habitat being destroyed by miners for a rare mineral called coltan. This mineral is apparently indispensable to the production of cell phones.

In a 2009 review of the usually irreversible harm EMFs cause, Johansson,[171] a neuroscientist, wrote that, "Today no-one would consider having a radio-active wrist watch with glowing digits (as you could in the 1950s), having your children's shoes fitted in a strong X-ray machine (as you could in the 1940s), keeping radium in open trays on your desk (as scientists could in the 1930s) or X-raying each other at your garden party (as physicians did in the 1920s). In retrospect that was just plain madness." However, the persons doing so were not less intelligent, however, knowledge was deficient.

Alterations in the Endocrine System

Most early studies, reviewed, for example, by WHO (1993) and later by Black and Heynick,[115] described thermally mediated responses of the endocrine system to RF smog exposure. Briefly, endocrine responses to acute RF (often CW 2.45 GHz) exposure are generally consistent with acute responses to EMF.[115] Also, chronic low-level exposures result in nonthermally mediated responses, which can cause not only hypersensitivity but arteriosclerosis, cancer, and neurovascular degenerative disease over time. Thyroid, adrenal, and pituitary changes have been found with EMF exposure, acutely and chronically.

Electromagnetic Field Alterations of the Eye

The lens of the eye is potentially sensitive to RF because it lacks a blood supply and so has a limited ability to dissipate heat. RF-induced cataracts are a well-established thermal effect of RF smog exposure in anesthetized rabbits; thresholds for prolonged (100–200 min) exposure lie between about 41° and 43°C, corresponding to localized SARs in the range 100–140 W kg^{-1}.[115] However, recent studies have confirmed anesthesia-restricted lenticular cooling through a reduction in local blood flow, thereby exacerbating the effects observed. Primates appear less susceptible to cataract induction than rabbits, and opacities have not been observed in them following either acute or prolonged exposures.

Studies from one laboratory suggesting that exposure of the eyes of anesthetized primates to pulsed RF could result in corneal lesions and vascular leakage from the iris were not corroborated by later studies by other groups using conscious primates. Transient changes were seen in the electroretinogram responses following exposure at high localized SARs, but the functional significance of this, if any, was not clear.[116]

The mortality of chicken embryos was found to increase to 75% from 16% in the control group after exposure to radiation from a GSM mobile phone.[117] This result is in agreement with the increased mortality of fertilized chicken eggs that was recorded after irradiation by low-power 9.152 GHz pulsed and continuous-wave microwaves.[118]

Electromagnetic Field Alterations of the Brain
Brain Tissue Removal and Homogenization of Mobile Phone–Exposed Rats
Parts of the brain (frontal lobe, hippocampus, and cerebellum) were quickly separated, immediately frozen in liquid nitrogen,

and then stored at 80°C until sample processing for further manipulation. In this study, researchers examined the protein expression levels in different mouse brain regions after whole-body exposure of Balb/c mice separately to mobile phone and wireless digital enhanced cordless telecommunications/telephone (DECT) base EMR.

The exposure conditions as explicitly described had an impact on the differential protein expression of a large number of brain proteins as follows.

Hippocampus: Eleven proteins were upregulated, whereas another 11 were downregulated after animal exposure to a wireless DECT base, compared to the sham-exposed animals. In addition, 37 proteins were found upregulated and 33 downregulated after exposure of the animals to a mobile phone compared to the sham-exposed (Table 2.4).

Frontal lobe: Twelve proteins were upregulated and 11 proteins were downregulated after exposure of the animals to a wireless base. The mobile phone exposure caused 19 proteins to become upregulated and 18 proteins to be downregulated (Table 2.4).

Cerebellum: Eight proteins were upregulated and 10 proteins were downregulated after exposure of the animals to a wireless base, whereas 36 proteins were upregulated and 18 proteins were downregulated in the mobile phone–exposed animal group (Table 2.4).

Summarizing, it seems that the mobile phone has a higher impact on all three brain regions isolated and studied compared to

TABLE 2.4 Number of Differentially Expressed Proteins across Three Major Brain Regions following Long-Term EMR Exposure to Conventional Mobile Phone (M) and DECT Wireless Base (B)

Proteins	Hippocampus		Frontal Lobe		Cerebellum	
	B	**M**	**B**	**M**	**B**	**M**
Upregulated	11	37	12	19	8	36
Downregulated	11	33	11	18	10	18
Total number of proteins changed	22	70	23	37	18	54

the wireless DECT base in the specific frequencies and intensities used. Furthermore, it is interesting that approximately the same number of proteins become upregulated or downregulated for a given brain region, except the cerebellum, where the vast majority of affected proteins (36) are upregulated.

This is the first report not only on mouse brain proteome effects induced by EMF but also on three major regions, namely the hippocampus, cerebellum, and frontal lobe. Therefore, there is no reference baseline to compare the actual results. The closest reports, but at the gene level, from Slalford's-Belyaev's groups have analyzed expression changes first in the cerebellum (GSM 900 MHz)[119] and second in the hippocampus and cortex (GSM 1800 MHz)[120] in rats. *They found significant alterations after a single 2-h and 6-h exposure, respectively.*

In this work, the researchers investigated separately the effects of chronic (8-month) daily whole-body exposure of Balb/c mice to EMR from (a) a typical medium SAR-level mobile phone (MP) GSM 900 MHz (3 h per day) and (b) the base of a wireless DECT (8 h per day) on the proteome of brain tissues. They showed that a large number of proteins become overexpressed or downregulated in three selected brain regions, namely the frontal lobe, hippocampus, and cerebellum. Most of these changes occur in the hippocampus, whereas the majority of the changes were induced by MP. This first observation could be explained by the fact that there are more concentrated functions in the hippocampus compared to the other two regions and that the hippocampal region may be more active metabolically. There is also a possibility of the existence of SAR hot spots in the hippocampus formation relative to the other brain regions.[121,122] The second fact (MP≫B) may be explained by the higher SAR value of MP radiation, though the exposure duration was less (3 h vs. 8 h). As shown, the overexpression/downregulation profile of the 143 proteins in the three brain regions may be helpful in understanding the behavioral and physiological effects reported in humans of EMR on brain function, including blood–brain barrier disruption, memory malfunction, oxidative stress, and so on.

Finally, some proteins have been affected by radiation simultaneously in two brain regions, namely the hippocampus-cerebellum, hippocampus-frontal lobe, and frontal lobe-cerebellum (16, 7, and 10 proteins, respectively). These include ApoE (hippocampus-cerebellum, related to memory function), NFL (also hippocampus-cerebellum, related to neuronal integrity), and a number of mitochondrial and metabolic proteins (asparatate aminotransferase, glutamate dehydrogenase, and others) that could be related to the recent observations on human brain after 50 min cell phone exposure in which the nonthermal effects were associated with increased brain glucose metabolism in the region closest to the MP antenna.[123,124]

Considering some of the affected proteins, scientists note the following.

The impressive protein downregulation of the nerve growth factor glial maturation factor beta (GMF) (300-fold for DECT bases and 8-fold for mobile phones), which is considered an intracellular signal transduction regulator in astrocytes,[125] may have an effect on the maintenance of the nervous system. As mentioned by the same authors, since "overexpression of GMF leads to interactions between neural cells, astrocytes, microglia and oligodendrocytes," they speculate that severe downregulation induced by DECT and MP radiation may inhibit the normal function of these cells. In addition, since this protein causes differentiation of brain cells, stimulation of neural regeneration, and inhibition of proliferation of tumor cells, its decrease could perhaps lead in the long run to tumor induction. Immunoblotting in GMF confirmed the proteomics data in general.

Glial fibrillary acidic *protein* (GFAP) overexpression by 15-fold in both types of radiation is in line with other single-protein expression reports following MP exposure of animals and is indicative of glial intermediate filament overproduction. This may in turn cause neurotransmitter uptake dysfunction and induction of gliosis,[126] which is a key step toward the epidemiologically suggested brain tumor increase upon long use of mobile phones.[127,128] Glial cells support neurons, release growth factors, and remove debris after injury or neuronal death. Astrocytes help form the blood–brain barrier that prevents toxic substances circulating in the blood from entering the brain. It was proposed many years ago that overexpression of GFAP is the response of astrocytes when oxidative stress occurs,[129] which is being reported to take place in brain tissues after exposure of guinea pigs to mobile phone radiation.[130] Since GFAP is a sensitive biomarker for neurotoxicity, these findings may indicate neuronal tissue injury caused by EMR or probable injury to the blood–brain barrier, reported to be an effect of exposure.[131,132] Immunoblotting with anti-GFAP confirmed the proteomics data in general.

ApoE is a class of apolipoprotein found in chylomicron and low density lipoproteins (LDLs) that binds to a specific receptor on liver and peripheral cells. It has been studied for its role in several biological processes not directly related to lipoprotein transport, including Alzheimer's disease (AD), immunoregulation, and cognition. So, the overexpression in the cerebellum and hippocampus after mobile exposure might be related to the memory deficits reported by this group.[111,133,134] This is in agreement with the observation that ApoE4 knock-in mice exhibit an age-dependent decrease in hilar GABAergic interneurons correlated with the extent of learning and memory deficits, as found by the Morris water maze task.[135]

Synapsin-2 and syntaxin-1 overexpression by both EMF radiation types (MP and DECT) in the hippocampus may indicate a compensatory neuronal response to EMF radiation by making more synapses.

Synaptotamin levels in the hippocampus are in line with the above-mentioned GMF dramatic downregulation. This protein species is known to function as a calcium sensor in the regulation of neurotransmitter release and hormone secretion.

The significance of the present study results may be noticeable in relation to the epidemiological, clinical, and other experimental data reported so far concerning behavioral deficits and brain structural/functional alterations induced by EMF in rodents. Although at the epidemiological level, Schüz et al.[136] found as an outcome of the Interphone study no overall increased risk of glioma or meningioma observed among cellular phone users, for long-term cellular phone users, the same authors suggested that the results need to be reconsidered before firm conclusions can be drawn. In fact, recent data by Hardell's group have provided solid evidence for a long-term effect on brain tumors,[127,128] which might be supported by the protein expression changes found in their results. Along the same lines, reports dealing with EMF-induced brain networking dysfunction can be explained. For instance, in a clinical study with 41 volunteers participating, it was reported that an 890-MHz mobile phone–like signal alters the integrity of the human blood–brain and blood–cerebrospinal fluid barriers.[137] There is also a relationship of MP radiation to behavioral problems in prenatally exposed children.[138]

Membrane A's

The data using the cordless DECT base as a source of EMF may appear surprising due to the low SAR level, as deduced by measuring the field within the animal cage, approximately 20 mW/Kg, but one explanation could be the intensity windows effect.[122,139] Interestingly, Salford's work with rats, applying a similar low SAR value (0.6 and 60 mW/Kg) but using mobile phone radiation for just 2 h per week for 55 weeks, demonstrated significantly altered performance during an episodic-like memory test.[77]

It is well established that, in general, the primary action of EMF on living tissue involves an increase of reactive oxygen species (ROS),[140] as demonstrated in exposed sperm[141,142] and under continuous stress conditions in *Drosophila* flies.[143] ROS accumulation and induced oxidative stress may lead to a signal transduction pathway (ERK kinases),[144,145] whereas at the same time, ion channels are disturbed.[144,146] Heat shock proteins are activated[144,147] and conformational change of enzymes[59] takes place. Thus, on the basis of the literature data and the findings, an EMF-impact mechanism can roughly be proposed involving ROS formation followed by stress activation, which may lead to the overexpression of Henoch-Schönlein purpura (HSP) (Figure 2.4). Through this event, several indirect changes may occur that alter the physiology of the brain cells, including

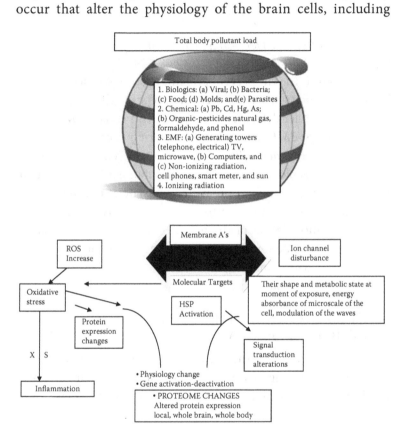

FIGURE 2.4 Schematic drawing depicting a suggested mechanism of EMF interaction with living matter. It is considered, on the basis of the available data and the present work, that the end result of protein expression changes may have been derived through a cascade of events starting from ROS increase and ion channel disturbance, followed by oxidative stress and signal transduction changers. A key role in the events may be played by heat shock protein activation.

DNA damage,[84] translation-transcription interference through protein conformation changes,[148] a possible cellular metabolism dysfunction, membrane dyspermeability,[149] and memory deficits.[133] It is clear that the effects of EMFs are very difficult to predict in the cells and that SAR values do not provide any information about the molecular mechanisms likely to take place during exposure.

Protein Activation by Electromagnetic Field Depends on Microenvironment Reception (Proteinomics)

The proteome is the entire set of proteins expressed by a genome. It is the set of expressed proteins in a given type of cell or organism at a given time under deformed conditions. This creates EMF smog.

Unlike drugs, EMFs are absorbed in a variety of different, diverse, and nonlinear ways depending on the "microenvironment" receiving the radiation. Table 2.5 shows (1) the orientation of the molecular targets, (2) their shape, (3) the metabolic state at the moment of exposure, (4) the energy absorbance at the microscale of the cell, and (5) the modulation of the waves. On this basis, it is rather difficult to replicate experiments under different conditions and cell systems, which may explain the discrepancy of the results among research groups.

ELECTROMAGNETIC FIELDS AND TOTAL BODY POLLUTANT LOAD

The results of triggering the proteome in the one area of the brain mean it can be triggered all over the brain and even the body, resulting in greater dysfunction in multiple systems (Figure 2.4;

TABLE 2.5   Ways EMF Is Absorbed

1. Orientation of molecular targets
2. Molecular target shape
3. Metabolic state at the moment of expression
4. Energy absorbance at the microscale of the cell
5. Modulation of the waves

i.e., the melatonin pituitary area) and increase antioxidant responses all over the body or different parts of a brain area like the hippocampus, the limbic system, and so on. This is especially so in the case of proteomics analysis used to approach the effects of EMFs. It is even more difficult since 2D electrophoresis (2DE) is inherently variable from one run to the next, especially when being performed in different labs. However, the 2DE approach is still largely refractory to high-throughput methods for a number of reasons and can be judiciously coupled to several types of biological experiments to provide meaningful data. Indeed, efforts to improve reproducibility have largely centered on robotics and improved visualization methods as better spot-detection algorithms,[150,151] and, as clearly stated by Ong and Pandey,[150] 2DE-based approaches can still be effectively used when applied with a clear understanding of its strengths and limitations.

The differential proteomic analysis results suggest that conventional MP and DECT base EMSs affect the proteome of the hippocampus, cerebellum, and frontal lobe following whole body exposure of Balb/c mice. Since this is the first report showing mouse brain proteome changes induced by EMFs, there is no reference baseline to which to compare the actual results. However, it is more likely that the observed proteome changes reflect EMF impact and not variability between individual mice, since it has been found just recently that genetic background in both outbred mouse stocks and inbred mouse strains has a negligible effect on the brain proteome profile.[152] Based on the currently available literature, it is assumed that EMF may function as a stress factor, creating ROSs and inducing oxidative stress, whereas at the same time, ion channels are disturbed and heat shock proteins are activated. This, in turn, may affect gene over/underexpression, possibly through transcription factor activation/deactivation[153] in a random manner, since EMF impact is nontargeted, although stress-related events within the cell are most likely affected. As an end result, functions related to the stress response may be triggered. The altered protein expression in this report may reflect such a cascade of events, in which some proteins are related to neural plasticity, whereas others belong to the general metabolic processes. The effects reported herein can be considered nonthermal since the actual SAR values calculated are well below the ICNIRP's (1998) guidelines. In any case, it is seriously considered by pioneers in the topic of EMFs that the relatively low field strengths capable of affecting biochemical reactions are a further indication that cells are in a position to sense potential danger long before there is an increase in temperature.[147]

Further work is underway to reveal the onset of proteome changes after short-term exposure conditions (data under analysis). Also, it is necessary to use multidisciplinary and multilevel approaches in order to delineate the mechanisms of EMF interaction with living organisms.

Advanced Metering Infrastructure and the Smart Energy Grid (Smart Meter)

Smart meters are another EMF generator that create EMF smog. Advanced metering infrastructure (AMI) technology is a key component of the smart energy grid. There are several purposes for going to an AMI "smart meter" infrastructure, including the following.

Reducing Operating Costs
Remote reading of meters would eliminate meter readers, allowing entities to save substantial costs in employee time and benefits, vehicle use, and gasoline costs. Smart meters can also be used to turn power on and off remotely, saving labor and travel costs when rentals become vacant or occupied. However, nothing is figured in about the adverse effects of smart meters, which have to be covered in the long-term costs.

Shifting Time of Use
Smart meters can measure and record total power usage for several intervals during the day. This will allow customers more for electrical usage at peak use hours, typically the early morning

(when people are getting up, taking showers, cooking breakfast) and late afternoon/early evening (when people return home from work, cook dinner, take showers, throw some clothes in the laundry, etc.). Time-of-use billing could create an incentive for customers to shift elective usage (laundry, recharging the electric car) away from peak usage hours.

Training Customers to Conserve Electricity
Smart meter technology can allow home owners to monitor their usage in real time over a home network with the meter. Direct feedback might encourage people to reduce their energy consumption.

MESH Network used in Smart Meters
MESH networking is a type of network topology used in which a device (i.e., smart meters) transmits its own data and also serves as a relay for other devices. From a biological point of view, AMI smart meters that transmit several times a minute can be considered an essentially constant source of RF exposure, thus creating a lot of EMF smog. Where these networks have been established in the last 2 years, large increases in reported acute symptoms have occurred. It is medically probable that this technology will be found to cause an increase in chronic health problems, including increased cancer, once sufficient time has passed for this to occur. Institute of Electrical and Electronics Engineers (IEEE) has already explored and tested a MESH option and chose to go forward without it. They want to ignore the complaints of those affected by smart meters even though they can be debilitating, causing weakness, fatigue, and brain dysfunction.

Solutions to Dampen Dirty Electricity
Fiber-Optic Communications
Fiber-optic communication between the utility and the house meter is an ideal solution from a health/environmental point of view, providing ample bandwidth without RF transmission. However, this technology would be quite expensive initially

to install, especially in parts of town where the power grid is underground. The cost might be prohibitive at this point in time until the long-term health effects are included in the equation. Like PLC, fiber optics would not communicate with the water meters.

Tower Communications Network (SENSUS)

The engineering system that is currently being considered is the SENSUS company's technology, where central towers communicate directly with meters on houses. This allows them to use more powerful radios on the smart meters, strong enough to communicate directly with a transmission tower without requiring that the message be passed from meter to meter across a MESH network. The community would be divided into about 13 zones, each of which would have a communication tower placed on an existing property within the zone, and these towers would communicate directly with house electric meters and radios on house water meters.

It is routine for utilities to collect data from these systems four times a day. But this routine was developed without consideration of the potential health risks of excessive RF transmission in the community. Usage of data does not need to be collected this frequently to achieve the main goals of the AMI program. From a practical point of view, the utility will continue to bill once a month and in theory could remotely collect that usage data once a month, minimizing the community's exposure to frequent and repetitive RF transmissions.

Transmission, even at this level of infrequency, would represent a minimal increase in the RF exposure to the community and would be unlikely to significantly increase the risk of chronic health problems in the community.

Each data transmission event would still be likely to provoke acute symptoms in individuals with EHS who lived near these transmission towers. But if these events occurred at an interval of once every two weeks or more, and at a predictable time of day, this might be a manageable level of exposure for those individuals.

In informal discussions with engineers, it has been said that they have looked into the issue of data collection frequency, and the longest that they could go between data collection events with the SENSUS system would be about three and a half days. This would appear to be a case where the technology has not been designed with an eye to minimizing RF transmission. Six daily time-of-use intervals times 30 days equals 180 intervals of usage data. If an iPod can store 64 gigabytes of music, it ought to be possible to give a smart meter enough memory to store 180 readings before transmitting them to the utility. Potential vendors could provide a meter with enough memory to store 2 to 4 weeks of data to enable the minimal RF footprint that is recommended.

Tower Communications and Water Meters
Water usage is billed once a month, and a single monthly reading of the meters would collect this data with minimal RF exposure to the community. Again, this data collection should occur in the daytime, not in the middle of the night.

Tower Communications and "Demand/Response"
From a public health perspective, the use of the system for "demand/response" load control is more problematic. As we understand it, a lot of this transmission would occur at night, when wind power production is high and demand is low. Towers would be transmitting every 15 minutes, to turn one cohort of water heaters on and another cohort off. The protocols required by the grid would require two-way communication with each meter in the cohort, acknowledging that house's participation in the cohort at that time.

This will involve a good deal of transmission in the system every 15 minutes, both from the towers potentially talking to hundreds of meters across the neighborhood and from the 2-watt radios on each house in the cohort talking back to the tower.

Communication of this frequency from the towers would be a significant additional layer of frequent nocturnal RF signal exposure to the residences within a few hundred meters of the towers.

If a great enough cohort of houses is involved, the transmissions from the meters on the houses could also increase the signal density in the residential areas enough to disrupt melatonin and sleep in a percentage of the population.[154]

This frequent level of activity in the demand/response system would be a significant additional RF burden on the community. It would make life in the residential area significantly more difficult for those individuals in the community who are currently already having acute problems. It would probably cause the onset of acute symptoms in a small percentage of the population who are not currently experiencing them. It would be likely to further increase the incidence of chronic adverse RF effects in the community.

PG&E's approach to the AMI rollout didn't involve a lot of public education. They just switched out the meters. Some people found that they were having trouble sleeping or experiencing headaches, ringing in the ears, vertigo, or other symptoms that hadn't been bothering them before. Soon the internet was awash in anecdotal reports and commentary about these adverse effects.

In response to a court order, PG&E provided documentation from the manufacturer of the meters that the average meter in the mesh network transmitted data signals to the utility 6 times a day, network management signals 15 times a day, timing signals 360 times a day, and beacon signals to the mesh network 9,600 times a day.[155] This penciled out to an average of roughly 7 transmissions a minute, 24 hours a day, coming out of every meter in the community.

By the end of 2011, multiple cities in California had either banned smart meters or placed a moratorium on their continued installation, and a lawsuit has been filed against PG&E with the California Public Utilities Commission.[156]

Elster MESH Advanced Metering Infrastructure Trial
In 2010, a trial of AMI infrastructure using the Elster REX2 smart meter occurred. Like the Silver Springs meter used by PG&E in California, the REX2 operates on a mesh network. The meters upload usage data to a central collection meter 4–6 times a day, but transmit short beacon signals to the network several times a minute.

Their website (Elster Solutions.com) states that these meters transmit "less than 10 seconds a day." But PG&E were unable to state how frequently transmissions actually occurred.

In January 2012, they used a Gigahertz Solutions HF35C analyzer to evaluate the output of one of these Elster meters in a residential neighborhood.

Background RF signals coming through the neighborhood were measured in a 360° circle around the monitoring position. The background RF averaged around 4 microwatts/square meter (uW/m^2), increasing to 8 or 10 uW/m^2 when they aimed the directional antenna at the radio towers at a distant cell phone tower.

The Elster meter's transmission rate was variable. In their observations, the meters are definitely transmitting several times a minute, sometimes four or five times a minute, and occasionally in bursts of significantly higher frequency.

At 5 feet from the smart meter, the peak strength of the beacon signal coming off the meter measured from 3800 to 11,000 uW/m^2. At 20 feet from the meter, the power density of the signal ranged from 362 to 493 uW/m^2, with occasional bursts at higher power output.

This means that at a distance of 20 feet, the power of the signal coming out of the Elster meter was about 100 times the power of the ambient background signal coming from any specific direction in the residential neighborhood.

This power density of 300+ to 400+ uW/m^2 was greater than the signal strength of the cell phone tower at 29th and Amazon, measured from about 200 meters away. So, filling a neighborhood with a mesh network of the Elster smart meters would be similar to placing every house in that neighborhood closer than 200 meters to a cell phone tower, each house constantly being pinged by the chatter of multiple beacon signals from the mesh. This was disconcerting, since recent research has shown that people living within 500 meters of a cell phone tower have increased incidence of headache, concentration difficulties, and sleep disorders, and also a significantly increased risk of some types of cancer.[22,114,157-159] When you put these facts together, it is not surprising that the installation of mesh smart meter networks in residential neighborhoods was followed by a surge of anecdotal evidence regarding headaches, insomnia, and other health complaints. From a medical perspective, based on a familiarity with current research in the biological effects of RF, this was a predictable consequence of PG&E's smart meter MESH network rollout.

Other Communication Options: Broadband Internet
We've been told that the power line communication option is not a feasible solution for demand/response control, since it lacks the bandwidth necessary for rapid communications between server and meters.

Serious consideration should be given to the potential use of broadband internet connections for demand/response communications. But in 2010, 82% of the households in one part of California had broadband internet connections, and this proportion continues to grow. Would it be technically possible to use these wired internet connections to communicate with the vast majority of the electric meters in the city, rather than building a new wireless infrastructure to do the job? Broadband internet communications would certainly have the bandwidth to do this, and a demand/response system is not expected to require the participation of every household in the community. If we acknowledge the health risks of RF communication (especially the robust nighttime communication expected for demand/response control), then an internet-based demand/response control system should be given serious consideration.

Regulatory Response

Regulations on exposure limits vary dramatically from country to country. In general, exposure limits have been mandated at a lower level in Russia and Eastern Europe, where research on the health effects of RF exposure has been performed for a longer period of time.[161]

The regulatory standards established by the FCC and the World Health Organization are based on defining safe levels against the thermal effects of RF (i.e., damage from being cooked by high levels of microwave exposure). The FCC has not established exposure standards for potential nonthermal or biological effects of microwave exposure,[162] even though there is ample evidence for its deleterious effects.

For example, the FCC has established limits for maximum permissible exposure (MPE). For the general population, the permissible level of exposure at 900 MHz is 600 uW/cm^2, and at 1800 MHz is 1000 uW/cm^2. These exposure levels were last updated in 1996 and are considered protective against thermal effects of microwave radiation. However, current scientific research shows that these are permissible levels of exposure.

The various forms of research described above have provided strong support for the fact that RF/EMF exposures can produce symptoms in human beings and that there is a percentage of the population that is more sensitive to this effect. Continued research suggests that this is not a static situation—that the prevalence of electrohypersensitivity is growing over time.

By the middle of the last decade, various government agencies were attempting to define the scope of the problem.[163]

The rollout of mobile phone technology occurred earlier in Scandinavia than in other places in the world, and governmental recognition of EHS as a health problem occurred earlier there than in other places. *By the year 2000, EHS was recognized as a disability by the Swedish government.*

In Stockholm, individuals with EHS can receive municipal support to reduce the presence and penetration of EMF/RF into their homes. The construction of a village with houses specifically designed to mitigate this problem is being considered. Patients with EHS have the legal right to receive mitigation in their workplace, and some hospitals have built low-EMF hospital rooms for use by such patients.[164]

Various government reports or reviews on the question of electrohypersensitivity have been commissioned in the last few years.[165,166] Legislation to address the problem has been proposed in some countries.[167,168] *Many libraries and schools in Europe have banned Wi-Fi due to concerns about health effects on employees and on the public. These levels are hundreds of times higher than the threshold levels for adverse "nonthermal" biological effects.*

For the past 10 years, the WHO has consistently equivocated on the issue of recognizing nonthermal biological effects from microwave RF exposure, despite the mounting research evidence of health problems and health risks produced by current levels of public exposure.

Table 2.6 shows exposure standards for various countries in 2001.[169] One can see that the United States shows higher exposures.

We are aware that the Soviet Union/Russia and the United States have spent vast fortunes since the early 1950s developing whole arsenals of EW weapons that, in many cases, use the same radio frequencies and the same—and even lower—power levels as do today's wireless devices, including baby monitors and smart meters.

Both the Russian and American militaries have known for at least 50 years that *low-level pulsed nonthermal radiation*—the kind emitted by all of today's wireless devices—is at least as harmful to humans, flora, and fauna as thermal radiation, possibly even more so. According to Flynn,[170] today the U.S. military, government, and industrial complex and its allies in Health Canada, Industry Canada, and other international agencies will not publicly admit that nonthermal radiation, especially pulsed, is harmful to humans, flora, and fauna, for fear that by doing so, it would jeopardize their military and commercial interests.

TABLE 2.6   Exposure Standards—Firstenberg 2001

Country	(uW/cm²)
New South Wales, Australia	0.001
Salzburg, Austria (for pulsed transmissions)	0.1
Russia	2–10
Bulgaria	2–10
Hungary	2–10
Switzerland	2–10
China	7–10
Italy	10
Auckland, New Zealand	50
Australia	200
New Zealand	200–1000
Japan	200–1000
Germany	200–1000
United States	200–1000
Canada	200–1000
United Kingdom	1000–10,000

Note: http://www.goodhealthinfo.net/radiation/radio_wave_
packet.pdf

Honest scientists (those not paid by industry or the military) know that:

- The human brain is exquisitely sensitive and can detect humanmade radiation as weak as one quadrillionth of a watt (0.000,000,000,000,001 W/cm²).

- The only safe level of *any* radiation is zero.

- All radiation is cumulative: it adds up from all emitting devices, layer upon layer.

- The damage done by cell phones to the human brain is also cumulative; that is, the total number of minutes spent in all of a person's cell phone calls adds up.

Of all of today's wireless devices, easily the most dangerous are wireless smart meters because of the aggregate amount of toxic radiation they generate in what utilities call "meshed-networked" communities, ranging in size from 500 to 5000 homes. Each home's smart meter contains two pulsed nonthermal microwave transmitters. This alone means that even the smallest community of just 500 homes will suddenly have 1000 new transmitters spewing toxic radiation.

While utilities seldom, if ever, mention it, homes in the future are expected to have 15 or so "smart" appliances—each of which will have its own pulsed microwave transmitter, receiver, and antenna circuit. In terms of just "smart" appliances, the smallest community of only 500 homes will have 7500 or so smart appliance transmitters, all of which will spew their toxic nonthermal radiation periodically throughout the 24-hour day in perpetuity. Note: it is predicted that the market for "smart" appliances will grow to USD 34.9 billion by 2020. From the above, it can easily be seen that even the smallest community (500 homes) will have in excess of 8500 new pulsing microwave transmitters, receivers, and antenna—none of which existed before smart meters were installed. In the largest communities of 5000 homes, there will be 10 times as many transmitters—85,000, all emitting their toxic radiation in perpetuity.[163] This mind-boggling amount of "new" radiation must then be added to whatever ambient level of radiation was already in a community before smart meters arrived.

- Dr. Lennart Hardell, PhD, arguably the world's most respected scientist today in cell phone radiation research, and his world-renowned Hardell Group of Sweden published five more studies in 2013—showing for the first time ever the harmful cumulative effects of people who used cell phones for 20 years. Such were his results that Dr. Hardell has called on the WHO to immediately upgrade radio frequency nonthermal radiation from Group 2B ("Possible") carcinogen to Group 1 ("Known") human carcinogen. If his recommendation is accepted, it means that the radiation emitted by all of today's wireless devices, including smart meters, baby monitors, cell phones, and so on, would be placed in the same carcinogenic category as tobacco and asbestos.

- Health Canada's Radiation Protection Bureau (who establish "Safety Code 6," Canada's so-called radiation "safety" exposure limits), their counterparts in Industry Canada, and the four "standard-setting" agencies of the world (WHO, ICNIRP, IEEE/Internal Committee on Electromagnetic Safety [ICES], and FCC) are all known to be nakedly corrupt. Even U.S. President Dwight D. Eisenhower warned the American people in 1961 to be on guard against a corrupt U.S. military, government, and industrial complex. Today's reality far surpasses his worst fears.

- Safety Code 6's radiation exposure limits are virtually identical to those of the WHO, ICNIRP, IEEE/ICES, and FCC, all being among the highest (most dangerous) in the world. Russia, China, Italy, Switzerland, and other countries have "safety" limits that are hundreds of times lower/safer than Canada's.

- The *BioInitiative 2007 Report* called for all countries of the world to reduce their radiation exposure limits to protect the public. Had Canada complied, Safety Code 6 would have had to be reduced 10,000 times.

- The *Seletun Scientific Statement,* written in 2011 by seven acclaimed scientists from five countries, warned that the global population is at risk, as there is serious disruption to biological systems now occurring.

- The *BioInitiative 2012 Report,* authored by 29 scientists, including 21 PhDs and 10 MD or MSc medical specialists from 10 countries, called for all countries of the world to reduce their radiation exposure limits to protect the public. Had Canada complied, Safety Code 6 would have had to be reduced by an astonishing 3–6 million times.

According to Flynn,[170] the following are withheld:

- Health Canada, B.C.'s provincial health officer (PHO) and their bedmates in the four "standard-setting" agencies have historically refused to admit and continue to deny that nonthermal radiation (the kind emitted by all of today's devices) is harmful to humans, as well as to flora and fauna.

- Health Canada, B.C.'s PHO and their aforementioned bedmates still will not admit that extremely low-frequency or power line 60-Hz electric and magnetic fields emitted by all electrical devices such as household electricity, household appliances, power tools, machinery, high-voltage transmission and distribution lines, and electrical substations are harmful to humans, flora, and fauna. Consequently, Canada, like the United States, has virtually *no* upper limit to which the public may be exposed—even 1000 mG (milli-Gauss) is allowed. By comparison, Russia allows its citizens to be subjected to EMFs not higher than 1.5 mG; Sweden, the country with the second-lowest (safest) limits, does not allow its citizens to be exposed to EMFs higher than 2.5 mG. Scientists have known for years that continuous exposure to EMFs of just 4 mG causes leukemia in children.

- Health Canada, B.C.'s PHO and their same bedmates collectively refuse to admit that there is such a thing as electrohypersensitivity. First identified by German scientists in 1932, countries in Europe began treating people suffering from EHS in the 1950s. Even the U.S. Air Force first reported it in 1955, and it has been treated in the United States for over 40 years. Even the former two-time prime minister of Norway, who became the director general of the WHO, Dr. Gro Harlem Brundtland, MD, MPH (Harvard) admitted in 2002 that she too suffers from EHS. Honest scientists know that in every country of the world, at least 3%–10% of the entire population already suffers from EHS, a number that some say will reach 50% by 2020.

Like electric utilities and wireless and telecom companies throughout North America, BC Hydro and Fortis BC have too easily allowed themselves to be drawn into the web of the enormously powerful U.S. military, government, and industrial complex. Today, this military-led American powerhouse promises other countries unprecedented profits and limitless business opportunities—if they agree to adopt the United States' dangerously high radiation exposure limits (which are actually determined by a powerful subcommittee within the IEEE). According to Flynn,[170] the principal members of this subcommittee are the U.S. Air Force, U.S. Army, Motorola, Nokia, Siemens, Alcatel-Lucent, and Bell.

It is also important to note that the same two doctors, William Bailey and Yakov Shkolnikov, who were retained by Fortis BC to defend smart meter technology to British Columbia Utilities Commission (BCUC) and the public in Fortis's application to roll out smart meters in 2013, are members of this committee,[170] as are several members of Health Canada and Industry Canada. Drs. Bailey and Shkolnikov are full-time apologists for the product defense firm Exponent Inc. Hence, one should now be able to understand that it is this U.S. military–led subcommittee within IEEE that establishes the United States' actual radiation exposure limits, which are then conveyed to the unsuspecting world through the offices of the WHO, ICNIRP, and FCC. Both Health Canada and Industry Canada have been in bed with the United States for many years. *Consequently, Safety Code 6's radiation exposure limits are virtually indistinguishable from those of the United States and are amongst the highest/most dangerous in the world.*

BACKGROUND

People, like all living things, are electrical beings; that is, we each have our own extremely delicate, exquisitely sensitive electrical system that operates at the cellular level. Easily the most vital, delicate, and complex organ of the body is the human brain, followed closely by the rest of the central and perfercial nervous and immune systems. Doctors can assess the electrical activity of the brain using an EEG, which measures the voltage fluctuations. By using *electrocardiography* (ECG), they can assess the electrical activity of the heart. Today's wireless consumer products and smart meters all emit *pulsed nonthermal* radio frequency radiation, which is foreign, disruptive, and harmful to us and all living things—especially to the brain—because living things don't recognize and cannot defend themselves against this humanmade, relatively much stronger *pulsed nonthermal* radiation. Never in the history of mankind has one technology— wireless radio—so completely captivated, enthralled, and helplessly addicted people of all ages, of all socioeconomic backgrounds, around the globe in so few years! No one can deny that wireless radio has radically altered our lives forever, making us far more efficient at work and at home, making life considerably easier and more enjoyable for most of us. Yet militaries have known since the 1950s that the electromagnetic energy ("radiation") emitted by the frequencies used in today's wireless devices belongs in weapons of war—not in consumer products and smart meters! (Pulsed nonthermal radiation from wireless devices is a global problem.)

As a former military electronic warfare (EW) specialist, Flynn[170] found it inconceivable that industry can be unaware that today's human-made radio frequencies are (and have been for many years) used in weapons of war. Surely they know that for at least 60 years, the Soviet/Russian and U.S. militaries have conducted exhaustive studies of the entire RF spectrum (3 kHz–300 GHz), meticulously cataloguing all frequencies in order to determine which are the most harmful to humans—especially to the human brain, central nervous system, and immune system. For more than 50 years, they've known that frequencies within the band 700 MHz–5 GHz penetrate all organ systems of the body, thus putting all organs at risk. Yet, this frequency band is precisely where all of today's wireless radio consumer products and smart meters have been assigned and authorized to operate (i.e., emit their pulsed nonthermal radiation). Even baby monitors work at this level. Militaries have known virtually all of the health hazards caused by wireless devices, including cancers, leukemia, cataracts, and so on for at least 50 years—but this has been kept secret by the U.S. military/government/industrial complex (MGIC) to preserve industrial profit.[170] (See Corrupt U.S. Military/Government/ Industrial Complex below).

All radio frequency and microwave frequency wireless devices emit invisible and virtually undetectable electromagnetic energy made up of electric and magnetic fields that travel at the speed of light. While EMF is the correct name, it is often also called EMR. Lower frequencies, such as household electricity (60 Hz) also have electric and magnetic fields that can be just as harmful to humans and other living things as RF and MW radiation.

Historical Facts

For 25 years (1953–1978), the Soviets irradiated the U.S. embassy in Moscow—but for only 6–8 hours a day, just five days a week.[170] The Soviets used the same microwave frequencies with similar modulation characteristics and, at times, with even much weaker signal strength than Industry Canada (IC) has authorized industry to use in today's baby monitors, smart meters, cell phones, cell towers, cordless phones, Wi-Fi routers, Wi-Fi and WiMAX zones, Bluetooth, tablet and laptop computers, and so on. Note: even though the embassy was irradiated for just 25% of the time, two consecutive U.S. ambassadors died of cancer, a third developed leukemia-like symptoms (bleeding of the eyes) from which he too died, and at least 16 women developed breast cancer; others developed leukemia and/or experienced a variety of illnesses. Today, we are all being constantly exposed 24/7/365 in perpetuity to low-level, pulsed, nonthermal microwave frequency radiation (pulsed nonthermal radiation) from a constellation of wireless devices whose radiation (labeled electrosmog by the WHO) is invisible and undetectable to an unsuspecting, sublimely addicted, and utterly uncaring and defenseless public.

Some efficient recognition started by 1961 when U.S. President Dwight D. Eisenhower, in his farewell speech to the nation, warned the American people to be on guard against the emergence of a corrupt U.S. military, government and industrial complex.[163]

1963—U.S. President John F. Kennedy said in a speech made to Columbia University on November 12: "The high office of the President has been used to foment a plot to destroy the American's freedom and before I leave office, I must inform the citizens of this plight." In 1969, the "Richmond Conference" (the final of 11 large conferences in the United States from 1955 to 1969) to determine the "Biologic Effects and Damages to Health Caused by Microwaves" presented overwhelming evidence of gastric bleeding, leukemia, chromosome breakages, cancer, and clouding of the eye lenses. It stressed that "brain functions were especially sensitive to RF EMF."[170]

1971—At least 15 years before any wireless radio devices appeared in North America—U.S. President Richard Nixon's own EMR Management Advisory Committee warned the White House that: "power (i.e. radiation) levels in and around American cities, airports and homes may already be biologically significant. The population at risk may well be the entire population. The consequences of undervaluing or misjudging the biological effects of long-term, low-level exposure could become a critical problem for the public health, especially if genetic effects are involved."[170]

1973—Evidence that the so-called radiation regulatory agencies, like Health Canada, were not protecting the public from nonthermal radiation can be seen in U.S. President Lyndon B. Johnson's State of the Union Address, in which he promised "to protect Americans against EMR—from TVs and other electronic equipment." (The wireless revolution was yet to occur.)

1973—Health Canada knew that microwave radiation is an "environmental pollutant" and a "threat to human health," but suppressed it. Evidence can be seen in the federal report LTR-CS-98 of April 1973. Yet Health Canada established its Code 6 by following ICNIRP guidelines for radio frequencies and publicly repeated the propaganda about nonthermal radiation being safe as recently as September 2010.[170]

1977—Still almost a decade before microwave-emitting wireless devices began flooding North America, Brodeur wrote his best-seller, *The Zapping of America*, which exposed the truth about the corruption surrounding microwave radiation. On the cover of his book, it states: "Microwave radiation can blind you, alter your behavior, cause genetic damage, even kill you. The risks have been hidden from you by the Pentagon, the State Department, and the electronics industry. With this book, the microwave cover-up is ended."

1979—Both the U.S. National Institute of Occupational Safety and Health (NIOSH) and Occupational Safety and Health Agency (OSHA) spoke out, saying they felt there was sufficient evidence that nonthermal radiation was harmful to people to warrant instituting the "precautionary principle"—but were ignored.[170]

1984—Steneck of the University of Michigan stated in his book *The Microwave Debate*: "The public have deliberately been deceived and kept ignorant. The effects of low-level exposure to RF radiation that have been withheld from the public extend from heart disease and cataracts to birth defects and cancer. Information about such hazards has been withheld to protect the military and its industrial contractors from the economic burden of lower standards."

1985—After parents brought suit, a Texas court ruled that Houston Lighting & Power had shown "callous disregard" of their children's health by siting a 345-kV line within 200 feet of a school and playground. The court ordered the utility to relocate the line.[170]

1986—The U.S. Environmental Protection Agency (EPA) urged the White House to classify all pulsed nonthermal radiation a possible carcinogen, but was ignored.[170]

1988—Brain tumors and EMFs. Workers exposed to 60-Hz fields in electric power utilities had an incidence of brain tumors 13 times greater than that in a comparable unexposed group.[170]

1989—Pulsed EMFs caused cancer in a study of electric utility workers in Quebec (follow-up, 1970–1988) and France (follow-up,

1978–1989): 2679 cases of cancer were identified. (Hydro Quebec was furious that the results were published and stopped further research.)[170]

1989—A Florida judge ruled that children could not play in a Boca Raton school yard that bordered high-voltage power lines. The suit was brought by three local parents who sought to close the Sandpiper Shores School because of potential electromagnetic field health hazards.

1990—The U.S. EPA admitted that the threat was real. It issued a draft report recommending that power-line (60 Hz) EMFs be classified a probable carcinogen. Industry pressure forced EPA to remove the recommendation.[170]

1993—U.S. Food and Drug Administration (FDA) data "strongly suggest" microwaves can promote cancer. FDA biologists concluded that the available data "strongly suggest" that microwaves can "accelerate the development of cancer." [170]

1994—The U.S. Air Force's own study confirmed existence of nonthermal EMF effects, including alterations to the central nervous system and cardiovascular system.[170] (Yet today, the United States, Canada, and their allies say EMFs are harmless.)

1994—Quebec Hydro. A McGill University study found up to a 10-fold increased risk of developing lung cancer among highly exposed utility workers with a "very clear" exposure-response relationship. Hydro-Quebec refused to allow further analysis and buried the report.[170]

1995—U.S. President Bill Clinton issued a formal memorandum stating that cell phone towers should not be placed at schools or in residential areas.[170]

1996—Ontario Hydro. Utility workers exposed to high levels of magnetic and electric fields had 11 times the expected rate of leukemia, with electrical fields being the more dominant. [170]

1999—The U.S. Consumer Affairs Commission said: "Current thermal guidelines associated with EMR are irrelevant. Cancer and Alzheimer's are associated with nonthermal EMR effects."[170]

2001—The WHO's own International Agency for Research on Cancer (IARC) said ELF EMFs are "possible carcinogens."

Twenty-one scientists from 10 countries unanimously found ELF EMFs are a Class 2B "Possible" carcinogen (electrical substations, power lines, house wiring, electrical appliances).[170]

2002—The Freiburger Appeal, signed by more than 6000 German doctors, listed 13 severe chronic illnesses and various disorders involving behavior, blood, heart, cancers, tinnitus, migraines, susceptibility to infections, and sleeplessness, all of which they ascribed to pulsed microwave mobile communications devices.[170]

2002—The FCC (which has no scientists of its own) set microwave radiation levels more than 10,000 times *higher* than levels that, according to the EPA, were causing illnesses all over the world.[170]

2002—Household electricity. Washington State Department of Health scientists reported that childhood leukemia is closely linked to household electricity—as are brain cancer, Lou Gehrig's disease, and miscarriage.[170]

2004—U.S. Senator John McCain, Chairman of the Senate Commerce Committee, said: "We have seen compelling evidence that there is an incestuous relationship between the defense industry and defense officials that is not good for America."[170]

2004—Nine different EMF studies from nine countries caused leading epidemiologists to see childhood leukemia risk at 4 mG. (Today scientists say an ambient level of <1 mG is safe, whereas Health Canada (HC) allows Canadians to be exposed to 1000 mG.)

2008—Former U.S. President Bill Clinton said, "Sarah, there's a government inside the government, and I don't control it," as quoted by senior White House reporter Sarah McClendon in reply to why he wasn't doing anything about UFO disclosure.

2010—U.S. President Barack Obama's own Cancer Advisory Panel urged the use of the precautionary principle for EMF. It identified the question of children's and adolescents' risk from using wireless devices as the most pressing issue.[170]

2010—The U.S. EPA publicly acknowledged that ELF EMFs are a serious threat to health. "More and more people living near HV lines have reported severe symptoms and even life-threatening diseases."[170]

2011—The Seletun Scientific Panel said that the global population is at risk and urged all countries to adopt new biologically based radiation limits.[170]

May, 2011—The Council of Europe (47 countries, 800 million people, who historically had followed ICNIRP's recommended radiation exposure limits) issued a news release urging all 47 governments to "reconsider" ICNIRP's exposure limits (which are followed by the WHO and Health Canada), "which have serious limitations and apply 'as low as reasonably achievable' (ALARA) principles ... Governments should take all reasonable measures' to reduce exposure to EMFs."[170]

June, 2011—The European Parliaments (47 countries, 800 million people) announced that it recognized for the first time the biological effects of EMFs on plants, animals, and human beings and said: "The need now is to protect citizens from EMR, particularly 'pregnant women, newborn babies and children.'"[170]

November, 2011—National radiation advisory authorities in Austria, Belgium, France, Germany, Russia, and Sweden recommended measures to minimize exposure to their citizens. (Canada and the United States remained silent.)[170]

2012—The BioInitiative Report concluded: (1) magnetic fields from ELF (60 Hz electricity) should be classified a Group 1 or "known carcinogen" and (2) radio frequency electromagnetic fields should be classified a "human carcinogen" (Group 1).[160]

2014—Leading scientists the world over considered that today's pulsed nonthermal radiation poses the biggest single threat to human health in mankind's entire history,[45-47] saying that: "We are all in an unprecedented situation where no one on earth lives outside the laboratory of this experiment. We are all guinea pigs in the hands of industry and government."

March, 2014—The Centers for Disease Control (U.S. Department of Health) estimated 1 in 68 children (1 in 42 boys versus 1 in 189 girls) had been identified with autism spectrum disorder (a 30% increase since 2012).[170]

July, 2014—Fifty-two world-class scientists from many countries condemned Health Canada's Safety Code 6, saying it is "flawed, and obsolete. They called on Health Canada to intervene to prevent an emerging public health crisis."[170]

August, 2014—Cell phone radiation in the United States. Ninety-eight scientists from 23 nations demanded stronger regulation of cell phone radiation. Of the 950 submissions to FCC regarding its 18-year old outmoded cell phone radiation regulations, almost all cited research that finds cell phone radiation harmful to humans.[50]

May, 2015—Scientists were so concerned for the future of the human race and all life forms on Earth that at least 206 of the world's most esteemed EMF scientists and cancer specialists from 40 countries launched the International EMF Scientist Appeal. Addressed to the secretary general of the United Nations, all UN member countries, and the director general of the WHO, it appealed to them all to "protect humans and wildlife from EMFs and wireless technology."[170]

August, 2015—One hundred new studies showed that using a cell phone for 20 minutes each day for 5 years increases the risk of one type of brain tumor by 300%, and talking on a cell phone for an hour a day for 4 years increases the risk of some tumors up to 500%.[170]

EMFs—Both fields are associated with all electricity, including electrical home appliances, tools, and machinery (when turned on), with overhead high voltage and power distribution lines, electrical substations, voltage transformers, and so on. Epidemiological studies showed that magnetic fields of just 3–4 mG can cause leukemia in children. Scientists recommended ambient levels of >1 mG. Russia recommended homes should have an ambient magnetic field of just 1.5 mG; Sweden recommended just 2.5 mG. Contrast this with the ICNIRP, WHO, and HC, all of whom said there is "no conclusive evidence" showing that EMFs are harmful.

For short-term duration, ICNIRP and WHO said ambient levels of up to 2000 mG are safe for the public and up to 10,000 mG for occupational exposures. Yet this same WHO, in 2001, classified EMFs a "possible carcinogen." HC said ambient levels of magnetic fields of up to 1000 mG are safe, whereas the IEEE/ICES—who determine the standard for the United States) said ambient levels up to 9000 mG are safe.[170]

2015—RF EMR protection—Neither Canada nor the United States had biological exposure guidelines for long-term exposure to today's low-level, pulsed, nonthermal radio/microwave frequency radiation (emitted by all of today's wireless radio devices).[170]

THE U.S. MILITARY/GOVERNMENT/ INDUSTRIAL COMPLEX

With military-like planning, precision and execution, the U.S. MGIC has:

1. Subordinated all five of the so-called "Five-Eyes" (intelligence-gathering) nations: the United States, Great Britain, Canada, Australia, and New Zealand.

2. Informed its allies (above) not to talk about nonthermal effects of microwave radiation. If it did not heat the body, they were not to talk about it.

3. Influenced the world's four so-called radiation regulatory agencies, the WHO, ICNIRP, IEEE/ICES, and FCC, into adopting radiation exposure limits compatible with those of the United States—all of which recognize only *thermal* radiation.

4. Managed to overcome huge national opposition to have President Obama appoint the career-long lobbyist and political fundraiser Tom Wheeler chairman of the all-powerful FCC!

This is the agency that "officially" determines the "safe"' radiation exposure limits for the United States (the FCC simply adopts those of ICES).[170]

5. Controlled and/or stifled the mainstream news media.[61-63]

The Cellular Telecommunications Internet Association (CTIA)—which lobbies government agencies across the United States on behalf of its members—spent almost USD800 million lobbying in the year 2013-14. (Today's FCC chairman, Tom Wheeler, was president and CEO of CTIA at that time.) If you remove the telecom industry–funded research, then the weight of the evidence overwhelmingly shows that cell phones cause health problems, including but not limited to brain tumors, eye cancer, testicular cancer. salivary gland tumors, non-Hodgkin's lymphoma, and leukemia.[170]

Canadian Wireless Telecommunications Association

In 2008, Canada endeared itself immeasurably to the U.S. MGIC when the Canadian Wireless Telecommunications Association (CWTA) hired Bernard Lord, Queen's Counsel (QC), Order of New Brunswick (ONB), to be its new president and CEO. Like its U.S. counterpart CTIA, CWTA is the main lobbyist group in Canada that lobbies all levels of government across the country on behalf of its wireless and telecom member companies as well as companies that develop and produce products and services for the industry (e.g., smart meters). Mr. Lord is a former two-time premier of New Brunswick, the former commissioner of Prime Minister Stephen Harper's Bilingual Commission, and is now also the chairman of the Ontario Power Generation. One can't imagine a more desirable, powerful, or influential CEO given his unrivaled connections in all governments across the country. With his personal connections within both the prime minister's office and the Conservative Party of Canada itself, there is likely no government door in Canada that Mr. Lord cannot open.

Even though HC finally bowed to widespread condemnation of its SC6 and lowered its "safe" radiation exposure "guidelines" some 66 times in 2015, it still remains one of the highest/ most dangerous radiation exposure limits in the world. China, Russia, Italy, and Poland have exposure limits much lower (safer) than does HC. Switzerland's limit is lower still for pregnant women, newborn babies, children, the elderly, and the medically challenged. The Swiss doctors' group Physicians for the Environment recently called for their limit to be reduced to just a tiny fraction of the other countries. Luxembourg/Bulgaria are at just 2.4 µW/cm². Salzburg's Health Department in Austria recommends only 0.001 µW/cm² for outdoor and 0.0001 for indoor exposure—which is millions of times lower than SC6. Swiss Re and Lloyd's of London, both major international insurance underwriters, and leading insurance groups will not insure any company or product against health-related claims attributed to nonthermal radiation.

Today, there are more than 6000 publications dealing with the harmful effects of radio wave radiation, including a complete bibliography of more than 2000 scientific reports on nonthermal radiation damage compiled before 1970 that are available. They were declassified by the U.S. military in 1971.[170]

With the current published data now available the harmful effects in radio wave radiation the public:

1. Would have recommended to the BC government that it needed to establish its own version of SC6, one that reflected "safe" radiation exposure limits more in line with Europe and, in 2007, with the *BioInitiative Report*—which called for exposure limits that are 10,000 times lower than is SC6 today.[170]

2. Would not have allowed our current generation of dangerous wireless baby monitors, Wi-Fi routers, cell phones, cordless phones, and tablet and laptop computers in our homes,

schools, nursing homes, hospitals, public buildings and spaces, and so on.

3. On learning that the WHO had classified all EMR a Class 2B ("Possible") carcinogen in 2011, would immediately have urged the BC government to declare wireless smart meters illegal in BC and to replace all such meters already installed with harmless, inexpensive, and reliable analog meters.

4. Would long ago have distanced her/himself from HC and urged the BC government to recognize electrohypersensitivity as a legitimate physical impairment, as has Sweden. Many other countries have recognized EHS, including the WHO.

5. Would long ago have distanced her/himself from HC and acknowledged that EMFs that accompany all electricity and all devices that run on electricity are harmful to humans. She/he would be aware that noindustry scientists say that magnetic field levels of <1 mG are safe and that EMFs as low as 2–3 mG cause leukemia in children—and many other diseases.

6. Long ago would have urged the BC government not to allow electric utilities to install their high-voltage transmission and distribution lines in close proximity to schools; residential buildings; places of work, play, and worship; or wherever people congregate.

7. Would have recommended to the government that it needed to establish its own research and development capability—free from industry influence—to conduct its own independent studies of EMR, EMF, and EHS, while always being mindful of what other countries of the world are doing and/or have learned.

Today's pulsed nonthermal radiation, which is emitted by all contemporary wireless radio consumer products and smart meters, is far more insidious and pernicious to humans and other life forms than any other technology ever devised by man! It is an absolute tragedy that today every person and living thing in the world is a victim, indeed a guinea pig, participating in what scientists consider the largest unauthorized technological experiment ever conducted in human history—without either the permission or knowledge of the subjects. Ironically, those naïve and gullible governments who were/ are responsible for allowing this unthinkable crime against humanity to happen will one day realize that they themselves were victims.

Militaries of the world have known for more than 50 years that: (a) microwaves are the perfect weapon—for human senses cannot detect them; (b) frequencies within the range 700 MHz–5 GHz are the most harmful to humans and other life forms; (c) the precise frequencies that, when combined with certain pulse modulation characteristics and power densities, are most harmful to the brain, central nervous system, and immune system and can cause cancers, leukemia, cataracts, and so on. Any and all militaries would be shocked into utter disbelief to know that today's so-called "democratic" governments are knowingly, deliberately, and callously authorizing today's untested (for safety) consumer wireless products and smart meters to operate (i.e., to emit their pulsed nonthermal radiation) on today's frequencies—which are known to be the most lethal frequencies known to man.

As a Canadian and a former military EW specialist, trained and practiced in EW, Flynn[170] is disgusted, appalled, and incensed to know that his very own federal government's HC and IC are so corrupt, so evil, and so devoid of any semblance of human decency that they are able to callously turn their backs on Canadians in order to maintain their privileged position within the corrupt U.S. military, government, industrial complex.[170] As they have for at least the past 30 years, HC today stubbornly:

1. Refuses to admit that nonthermal radiation—especially when pulsed—can be harmful to people and other life forms.

2. Refuses to admit that ELF (power-line 60 Hz) electric and magnetic fields can be harmful to people and other life forms.

3. Refuses to admit that there is such a condition as electrohypersensitivity, saying that those claiming to have EHS suffer some idiopathic or psychiatric problem unrelated to wireless technology. (EHS was first identified by German scientists in 1932; the U.S. military first reported it in 1955, and many countries of the world—even the WHO—have since recognized it.)

4. Refuses to consider the hundreds of Soviet/Russian studies, the more than 2000 U.S. military studies, and the more than 6000 nonindustry peer-reviewed studies from every corner of the globe—all of which show, beyond any reasonable doubt, the harmful effects humans and other life forms experience when chronically exposed to low-level, pulsed, nonthermal RF and MW radiation, that is, the kind emitted by today's wireless consumer products and smart meters.

5. Refuses to heed the countless pleas, urgings, and warnings from Russia and various countries of the world—even the Council of Europe itself (47 countries, 800 million people)— who urge all countries of the world to lower their radiation exposure limits in order to protect pregnant women, children, the sick, and the elderly.

6. Refuses to note or heed that European countries began pulling Wi-Fi from schools, libraries, and so on many years ago; increasingly, others are banning cell phones and other wireless technology from schools, other public places, and so on.

7. Refuses to note or heed that many countries of the world have imposed the precautionary principle in order to protect their people.

8. Refuses to note or heed the alarms and protests that have been raised over the past 30 years or more by major U.S. federal agencies, such as the EPA, NIEHS, FDA, NIOSH,

OSHA, Consumer Affairs Commission, and so on, all of whom expressed concern about the high levels of radiation permitted by the United States and/or that the United States' exposure limits recognize only thermal radiation.

Smart Meter Health Effects Survey: Results, Analysis, and Report

Conrad's[81] survey was designed to discover if the health effects/ symptoms that many persons have been attributing to smart meter exposures were really caused by those exposures. The survey essentially collected testimonials of personal experiences with smart meters, broken down into answers to approximately 50 questions, most of them multiple choice. Since all questions required an answer, all respondents answered identical questions via a choice of identical answers. This provided uniformity of the data collected, enabling detailed analysis and comparison of their experiences.

Approximately 75% of respondents were from the United States and the rest from Canada and Australia.

Of the 210 respondents, 9 were PhDs, 42 MS or MA, 70 BS or BA, 1 MD, and 1 DDS.

Before smart meters, 23.3% (calculated from Q2a) of the 210 respondents considered themselves to have ES.

Now, after smart meters, 67.6% (Q32) of the 210 respondents consider themselves to have ES. Note that the majority of these (62.7%, calculated from Q32a) feel certain that their exposure to smart meters was responsible for initiating their ES.

Of the 49 persons who already considered themselves to have ES before SM exposure, all 49 (100%) felt that their exposure to SM made their ES not only worse, but "much worse."

Cell Phone, Computer, and Wi-Fi Use before and after Smart Meters

In order to ascertain the effect smart meter installation had on respondents' ability to use common electronic devices such as cell phones, Wi-Fi, and computers, we looked at device use before and after smart meters. We found very clear evidence that smart meter exposure adversely affected respondents' ability to use other RF devices without incurring harmful symptoms.[81] In the survey, computer use is addressed in questions 4 and 29, Wi-Fi in questions 5 and 30, and cell phones in questions 6 and 31.

Computer Use

Before smart meters, nearly 79% of respondents were using computers without symptoms, while about 20% were using computers despite having symptoms from computer use. Following smart meter exposure, those able to operate a computer without symptoms dropped (from 79%) to 39% (about one-half of before), while those showing symptoms from computer use nearly tripled (from 20%) to 57%.

Wi-Fi Use

Before smart meters, about 40% of respondents were using Wi-Fi without symptoms, 11% were using Wi-Fi but with symptoms from it, and 17% were not using Wi-Fi because it had caused symptoms in the past. Following smart meter exposure, those able to use Wi-Fi without symptoms dropped (from 40%) to 18% (less than one-half of before), while those continuing to use Wi-Fi but with symptoms from it nearly tripled (from 11%) to 28%. The number of respondents who could not use Wi-Fi at all because of symptoms more than doubled (from 17%) to 41%.

Cell Phone Use

Before smart meters, 50% of respondents were using cell phones without symptoms, while 18% used cell phones but with symptoms. Fourteen percent of respondents did not use cell phones because of symptoms. Following smart meters, those able to use cell phones without symptoms dropped (from 50%) to 24% (about one-half of before), and those with symptoms from cell phone use more than doubled (from 18%) to 39%. After smart meters, those who did not use cell phones at all because of symptoms nearly doubled (from 14%) to 26%.

Obviously, the inability to use these modern tools severely inhibits our respondents in their personal and economic lives. Their ability to live normal lives in the twenty-first century has been severely compromised. This change in ability to use these devices is directly correlated to smart meter exposure.[81]

Symptoms, New and Worsened, Correlated to Smart Meter Exposure

The survey asks respondents to identify, from a list of 21 symptoms, the ones they found to be associated with their smart meter exposure. They also allowed space for respondents to write in other symptoms they felt were associated with smart meters. Specifically they wanted to know *what* their symptoms were, their *intensity*, and which symptoms were *new* (never before experienced) to them since smart meters and which ones were previously experienced symptoms that were *worsened* by smart meter exposure.

They found 14 symptoms that many individuals suffered in common, including ear ringing, headaches, difficulty concentrating, insomnia, and heart arrhythmias. Many of these same symptoms have been cited in previous literature on low-level RF and, for example, in the EMF Safety Network Survey (also submitted in our public utilities commission [PUC] testimony). How long can you go without sleep, how well can you respond to workplace and personal situations and stressors when having cognitive difficulties, how much pressure in your head must you tolerate? While any of these symptoms by themselves on occasion may be of little consequence, taken in combination and at a severe level, they are quite enough to force people from their homes and workplaces. Here there is no 10–30 year latency period as there may be from cell phone exposure to diagnosis of glioma; instead, the devastating results are essentially immediate.[81]

Almost one-third, 32.4%, of the 210 respondents were initially healthy, had no prior concerns about EMF from smart meters, and began to experience symptoms at a time (before discovery of SM or well after discovery of SM) indicating their symptoms did not develop due to any knowledge of or concern about smart meters. The number of respondents who considered themselves ES before smart meters and who began to experience their new/ worsened symptoms *before* they discovered the smart meter, totaled 19, or 9% of the 210 respondents.

The suffering and the social and economic effects of chronic debilitating symptoms victims have experienced since smart meter exposure simply cannot be ignored and provide ample evidence there is something about smart meters making them extremely harmful to at least some, and possibly eventually to all, persons. While there is obviously only a portion of our population consciously realizing and manifesting ES/EHS symptoms at present (the canaries), everyone is being exposed. No one knows whether they or a family member is predisposed toward developing electrical sensitivities. Predisposition does not depend on opinions, beliefs, background, or occupation. At this point, exposure to a smart meter is like playing Russian roulette.

REFERENCES

1. Durham, M. O. and R. A. Durham. 1995. Lightning, grounding and protection for control systems. *IEEE Trans. Ind. Appl.* 45–54.
2. Shulman, L. 1986. Electromagnetic interference and grounding consideration in distributed control systems. *IEEE Ind. Appl. Soc. Newsl.* May/June 1986.
3. Durham, M. O. and R. Strattan. 1990. *Harmonics on AC Power Systems. Frontiers of Power.* Stillwater, OK: Oklahoma State University.
4. Lightning/EMP and Grounding Solutions, PolyPhaser Com. 1992. Minden. NV.
5. Milham, S. and D. Stetzer. 2013. Dirty electricity, chronic stress, neurotransmitters and disease. *Electromagn Biol. Med.* 500–507.
6. Gottlieb, B. and the editors of *Prevention.* 2015. *Health-Defense: How to Stay Vibrantly Healthy in a Toxic World.* Rodale, Inc.
7. Segell, M. 2011. Is dirty electricity making you sick? *Prevention.* Retrieved 2015-12-1. http://www.prevention.com/health/healthy-living/electromagnetic-fields-and-your-health

8. ElectroSensitivity UK News. 2010. Retrieved 2015-12-1. http://www.es-uk.info/news/20100301_main_newsletter.pdf

9. Havas, M. http://electromagnetichealth.org/images/Heart-Irreglarities-Linked-to-Wireless-Radiation.png. Retrieved 2009-11-2, 2018-09-21. ElectromagneticHealth.org.

10. Milham, Jr. S. 1988. Ham Radio Operators' High Cancer Rate Poses a Puzzle. Retrieved 2018-09-19. http://articles.latimes.com/1988-01-03/news/mn-32536_1_radio-operator

11. Hocking. 1996. Can living next to a TV antenna transmitter be harmful? *Med J Austral* 165:601–605.

12. Milham, S. and L. L. Morgan. 2008. A new electromagnetic exposure metric: High frequency voltage transients associated with increased cancer incidence in teachers in a California school. *Am. J. Ind. Med.* 51(8):579–586.

13. Huttunen, P., O. Hänninen, and R. Myllylä. 2009. FM-radio and TV tower signals can cause spontaneous hand movements near moving RF reflector. *Pathophysiology* 16(2–3):201–204.

14. Philips, Alasdair. 2009. #1156: Is Electro Smog an Emerging Occupational Health Issue? Retrieved 2018-09-19. http://www.dailymail.co.uk/news/article-1229069/Is-electro-smog-causing-headache.html

15. O'Neill, B. P., N. J. Iturria, M. J. Link, M. J. Link, B. E. Pollock, K. V. Ballman, and J. R. O'Fallon. 2003. A comparison of surgical resection and stereotactic radiosurgery in the treatment of solitary brain metastases. *Int. J. Radiat. Oncol. Biol. Phys.* 55(5):1169–1176.

16. Panagopoulos, D. J. and L. H. Margaritis. 2008. Mobile telephony radiation effects on living organisms. In: Harper, A. C. and Buress, R. V. (Eds.). *Mobile Telephone.* Chapter 3. Nova Science Publishers, Inc., pp. 107–49.

17. Hillebrand, F. (Ed). 2001. *GSM and UTMS.* Wiley.

18. Clark, M. P. 2001. *Networks and Telecommunications,* 2nd ed. Wiley.

19. Hyland, G. J. 2000. Physics and biology of mobile telephony. *Lancet* 356:1833–1836.

20. Hamnerius, I. and T. Uddmar. 2000. Microwave exposure from mobile phones and base stations in Sweden. *Proceedings, International Conference on Cell Tower Siting,* Salzburg, pp. 52–63, www.land-salzburg.gv.at/celltower

21. Tisal, J. 1998. *GSM Cellular Radio Telephony.* West Sussex, England: J. Wiley & Sons.

22. Abdel-Rassoul, G., O. A. El-Fateh, M. A. Salem, A. Michael, F. Farahat, M. El-Batanouny, and E. Salem. 2007. Neurobehavioral effects among inhabitants around mobile phone base stations. *Neurotoxicology* 28(2):434–440.

23. Navarro Enrique, A., J. Segura, M. Portolés, and C. Gómez- Perretta de Mateo. 2003. The microwave syndrome: A preliminary study in Spain. *Electromagn Biol. Med.* 22(2–3):161–169.

24. Santini, M. T., A. Ferrante, G. Rainaldi, P. Indovina, and P. L. Indovina. 2005. Extremely low frequency (ELF) magnetic fields and apoptosis: A review. *Int. J. Radiat. Biol.*. 81(1):1–11.

25. Bortkiewicz, A., M. Zmyślony, A. Szyjkowska, and E. Gadzicka. 2004. Subjective symptoms reported by people living in the vicinity of cellular phone base stations: Review. *Med. Pr.*. 55(4):345–351. (In Polish.)

26. Hutter, H.-P., H. Moshammer, P. Wallner, and M. Kundi. 2006. Subjective symptoms, sleeping problems, and cognitive performance in subjects living near mobile phone base stations. *Occup. Environ. Med.* 63:307–313.

27. Banik, S., S. Bandyopadhyay, and S. Ganguly. 2003. Bioeffects of microwave-a brief review. *Bioresour. Technol.*. 87(2):155–159.

28. Diem, E., C. Schwarz, F. Adlkofer, O. Jahn, and H. Rudiger. 2005. Non-thermal DNA breakage by mobile-phone radiation (1800 MHz) in human fibroblasts and in transformed GFSH-R17 rat granulosa cells *in vitro. Mutat. Res.*. 583(2):178–183.

29. Panagopoulos, D. J., A. Karabarbounis, and L. H. Margaritis. 2004. Effect of GSM 900-MHz mobile phone radiation on the reproductive capacity of *Drosophila melanogaster. Electromagn Biol. Med.* 23(1):29–43.

30. Panagopoulos, D. J., E. D. Chavdoula, I. P. Nezis, and L. H. Margaritis. 2007. Cell death induced by GSM 900 MHz and DCS 1800 MHz mobile telephony radiation. *Mutat. Res.* 626:69–78.

31. Panagopoulos, D. J., E. D. Chavdoula, A. Karabarbounis, and L. H. Margaritis. 2007. Comparison of bioactivity between GSM 900 MHz and DCS 1800 MHz mobile telephony radiation. *Electromagn Biol. Med.* 26(1).

32. Leszczynski, D., S. Joenväärä, J. Reivinen, and R. Kuokka. 2002. Non-thermal activation of the hsp27/p38MAPK stress pathway by mobile phone radiation in human endothelial cells: Molecular mechanism for cancer- and blood-brain barrier-related effects. *Differentiation* 70(2–3):120–129.

33. Schirmacher, A., S. Winters, S. Fischer, J. Goeke, H. J. Galla, U. Kullnick, E. B. Ringelstein, and F. Stögbauer. 2000. Electromagnetic fields (1.8 GHz) increase the permeability to sucrose of the blood-brain barrier *in vitro*. *Bioelectromagnetics* 21(5):338–345.

34. Velizarov, S., P. Raskmark, and S. Kwee. 1999. The effects of radiofrequency fields on cell proliferation are non-thermal. *Bioelectrochemistry and Bioenergetics* 48:177–180.

35. Bawin, S. M., L. K. Kaczmarek, and W. R. Adey. 1975. Effects of modulated VMF fields, on the central nervous system. *Ann. NY Acad. Sci.* 247:74–81.

36. Bawin, S. M., W. R. Adey, and I. M. Sabbot. 1978. Ionic factors in release of 45Ca 2+ from chick cerebral tissue by electromagnetic fields. *Proc. Natl. Acad. Sci. USA* 75:6314–6318.

37. Blackman, C. F., S. G. Benane, J. A. Elder, D. E. House, J. A. Lampe, and J. M. Faulk. 1980. Induction of calcium–ion efflux from brain tissue by radiofrequency radiation: Effect of sample number and modulation frequency on the power—density window. *Bioelectromagnetics (NY)* 1:35–43.

38. Blackman, C. F., L. S. Kinney, D. E. House, and W. T. Joines. 1989. Multiple power-density windows and their possible origin. *Bioelectromagnetics* 10(2):115–128.

39. Ozguner, F., Y. Bardak, and S. Comlekci. 2006. Protective effects of melatonin and caffeic acid phenethyl ester against retinal oxidative stress in long-term use of mobile phone: A comparative study. *Mol. Cell. Biochem.* 282(1–2):83–88.

40. Oktem, F., F. Ozguner, H. Mollaoglu, A. Koyu, and E. Uz. 2005. Oxidative damage in the kidney induced by 900-MHz-emitted mobile phone: Protection by melatonin. *Arch. Med. Res.* 36(4):350–355.

41. Forgács, Z., G. Kubinyi, G. Sinay, J. Bakos, A. Hudák, A. Surján, C. Révész, and G. Thuróczy. 2005. Effects of 1800 MHz GSM-like exposure on the gonadal function and hematological parameters of male mice. *Magy. Onkol.* 49(2):149–151.

42. Repacholi, M. H., A. Basten, V. Gebski, D. Noonan, J. Finnie, A. W. Harris. 1997. Lymphomas in E mu-Pim1 transgenic mice exposed to pulsed 900 MHZ electromagnetic fields. *Radiat. Res.* 147(5):631–640.

43. Paulraj, R. and J. Behari. 2006. Protein kinase C activity in developing rat brain cells exposed to 2.45 GHz radiation. *Electromagn Biol. Med.* 25(1):61–70.

44. Paulraj, R. and J. Behari. 2006. Single strand DNA breaks in rat brain cells exposed to microwave radiation. *Mutat. Res.* 596(1–2):76–80.
45. Magras, I. N. and T. D. Xenos. 1997. RF radiation-induced changes in the prenatal development of mice. *Bioelectromagnetics* 18:455–461.
46. Pyrpasopoulou, A., V. Kotoula, A. Cheva, P. Hytiroglou, E. Nikolakaki, I. N. Magras, T. D. Xenos, T. D. Tsiboukis, and G. Karkavelas. 2004. Bone morphogenetic protein expression in newborn rat kidneys after prenatal exposure to radiofrequency radiation. *Bioelectromagnetics* 25(3):216–227.
47. Demsia, G., D. Vlastos, and D. P. Matthopoulos. 2004. Effect of 910-MHz electromagnetic field on rat bone marrow. *Sci. World J.* 4(Suppl 2):48–54.
48. Sommer, A. M., A. K. Bitz, J. Streckert, V. W. Hansen, and A. Lerchl. 2007. Lymphoma development in mice chronically exposed to UMTS-modulated radiofrequency electromagnetic fields. *Radiat. Res.* 168(1):72–80.
49. Oberto, G., K. Rolfo, P. Yu, M. Carbonatto, S. Peano, N. Kuster, S. Ebert, and S. Tofani. 2007. Carcinogenicity study of 217 Hz pulsed 900 MHz electromagnetic fields in Pim1 transgenic mice. *Radiat. Res.* 168(3):316–326.
50. Juutilainen, J., P. Heikkinen, H. Soikkeli, and J. Mäki-Paakkanen. 2007. Micronucleus frequency in erythrocytes of mice after long-term exposure to radiofrequency radiation. *Int. J. Radiat. Biol.* 83(4):213–220.
51. Tillmann, T., H. Ernst, S. Ebert, N. Kuster, W. Behnke, S. Rittinghausen, and C. Dasenbrock. 2007. Carcinogenicity study of GSM and DCS wireless communication signals in B6C3F1 mice. *Bioelectromagnetics* 28(3):173–187.
52. Gatta, L., R. Pinto, V. Ubaldi, L. Pace, P. Galloni, G. A. Lovisolo, C. Marino, and C. Pioli. 2003. Effects of *in vivo* exposure to GSM-modulated 900 MHz radiation on mouse peripheral lymphocytes. *Radiat. Res.* 160(5):600–605.
53. Hennies, K., H. P. Neitzke, and H. Voigt. 2000. *Mobile Telecommunications and Health. Review of the Current Scientific Research in View of Precautionary Health Protection.* ECOLOG-Institut.
54. Lixia, S., K. Yao, W. Kaijun, L. Deqiang, H. Huajun, G. Xiangwei, W. Baohong, Z. Wei, L. Jianling, and W. Wei. 2006. Effects of 1.8 GHz radiofrequency field on DNA damage and expression of heat shock protein 70 in human lens epithelial cells. *Mutat. Res.* 602(1–2):135–142.

55. Remondini, D., R. Nylund, J. Reivinen et al. 2006. Gene expression changes in human cells after exposure to mobile phone microwaves. *Proteomics* 6(17):4745–4754.

56. Nylund, R. and D. Leszczynski. 2006. Mobile phone radiation causes changes in gene and protein expression in human endothelial cell lines and the response seems to be genome- and proteome-dependent. *Proteomics* 6(17):4769–4780.

57. Weisbrot, D., H. Lin, L. Ye, M. Blank, and R. Goodman. 2003. Effects of mobile phone radiation on reproduction and development in *Drosophila melanogaster. J. Cell. Biochem.* 89(1):48–55.

58. French, P. W., R. Penny, J. A. Laurence, and D. R. McKenzie. 2001. Mobile phones, heat shock proteins and cancer. *Differentiation* 67(4–5):93–97.

59. Barteri, M., A. Pala, and S. Rotella. 2005. Structural and kinetic effects of mobile phone microwaves on acetylcholinesterase activity. *Biophys. Chem.* 113(3):245–253.

60. Mancinelli, F., M. Caraglia, A. Abbruzzese, G. d'Ambrosio, R. Massa, and E. Bismuto. 2004. Nonthermal effects of electromagnetic fields at mobile phone frequency on the refolding of an intracellular protein: Myoglobin. *J. Cell. Biochem.* 93(1):188–196.

61. Pacini, S., M. Ruggiero, I. Sardi, S. Aterini, F. Gulisano, and M. Gulisano. 2002. Exposure to global system for mobile communication (GSM) cellular phone radiofrequency alters gene expression, proliferation, and morphology of human skin fibroblasts. *Oncol. Res.* 13(1):19–24.

62. Kwee, S. and P. Raskmark. 1998. Changes in cell proliferation due to environmental nonionizing radiation: 2. Microwave radiation. *Bioelectrochemistry and Bioenergetics* 44:251–255.

63. Mashevich, M., D. Folkman, A. Kesar, A. Barbul, R. Korenstein, E. Jerby, and L. Avivi. 2003. Exposure of human peripheral blood lymphocytes to electromagnetic fields associated with cellular phones leads to chromosomal instability. *Bioelectromagnetics* 24(2):82–90.

64. Krause, C. M., L. Sillanmäki, M. Koivisto, A. Häggqvist, C. Saarela, A. Revonsuo, M. Laine, and H. Hämäläinen. 2000. Effects of electromagnetic fields emitted by cellular phones on the electroencephalogram during a visual working memory task. *Int. J. Radiat. Biol.* 76(12):1659–1667.

65. Huber, R., V. Treyer, A. A. Borbely et al. 2002. Electromagnetic fields, such as those from mobile phones, alter regional cerebral blood flow and sleep and waking EEG. *J. Sleep Res.* 11(4):289–295.

66. Loughran, S. P., A. W. Wood, J. M. Barton, R. J. Croft, B. Thompson, and C. Stough. 2005. The effect of electromagnetic fields emitted by mobile phones on human sleep. *Neuroreport* 16(17):1973–1976.

67. Röschke J. and K. Mann. 1997. No short-term effects of digital mobile radio telephone on the awake human electroencephalogram. *Bioelectromagnetics* 18:172–176.

68. Wagner, P., J. Roschke, K. Mann, W. Hiller, and C. Frank. 1998. Human sleep under the influence of pulsed radiofrequency electromagnetic fields: A polysomnographic study using standardized conditions. *Bioelectromagnetics* 19:199–202.

69. Esen, F. and H. Esen. 2006. Effect of electromagnetic fields emitted by cellular phones on the latency of evoked electrodermal activity. *Int. J. Neurosci.* 116(3):321–329.

70. Gadhia, P. K., T. Shah, A. Mistry, M. Pithawala, and D. Tamakuwala. 2003. A preliminary study to assess possible chromosomal damage among users of digital mobile phones. *Electromagn Biol. Med.* 22(2):149–159.

71. Tahvanainen, K., J. Nino, P. Halonen, T. Kuusela, T. Laitinen, E. Lansimies, J. Hartikainen, M. Hietanen, and H. Lindholm. 2004. Cellular phone use does not acutely affect blood pressure or heart rate of humans. *Bioelectromagnetics* 25(2):73–83.

72. Burch, J. B., J. S. Reif, C. W. Noonan, T. Ichinose, A. M. Bachand, T. L. Koleber, and M. G. Yost. 2002. Melatonin metabolite excretion among cellular telephone users. *Int. J. Radiat. Biol.* 78(11):1029–1036.

73. Zwamborn, A. P. M., S. H. J. A. Vossen, B. J. A. M. van Leersum, M. A. Ouwens, and W. N. Mäkel. 2003. *Effects of Global Communication System Radio-Frequency Fields on Well-Being and Cognitive Functions of Human Subjects with and without Subjective Complaints.* FEL-03-C148. The Hague, the Netherlands: TNO Physics and Electronics Laboratory. Available: http://home.tiscali. be/milieugezondheid/dossiers/gsm/TNO_rapport_Nederland_ sept_2003.pdf

74. Regel, S. J., S. Negovetic, M. Röösli, V. Berdiñas, J. Schuderer, A. Huss, U. Lott, N. Kuster, and P. Achermann. 2006. UMTS base station-like exposure, well-being, and cognitive performance. *Environ. Health Perspect.* 114(8):1270–1275.

75. Hardell, L., M. Carlberg, F. Söderqvist, K. H. Mild, and L. L. Morgan. 2007. Long-term use of cellular phones and brain tumours: Increased risk associated with use for > or =10 years. *Occup. Environ. Med.* 64(9):626–632. Review.

76. Hardell, L., M. Carlberg, and K. Hansson Mild. 2006. Pooled analysis of two case-control studies on use of cellular and cordless telephones and the risk for malignant brain tumours diagnosed in 1997–2003. *Int. Arch. Occup. Environ. Health* 79(8):630–639.

77. Salford, L. G., A. Brun, K. Sturesson, J. L. Eberhardt, and B. R. Persson. 1994. Permeability of the blood-brain barrier induced by 915 MHz electromagnetic radiation, continuous wave and modulated at 8, 16, 50, and 200 Hz. *Microsc. Res. Tech.* 27(6):535–542.

78. Oscar, K. J. and T. D. Hawkins. 1977. Microwave alteration of the blood-brain barrier system of rats. *Brain Res.* 126:281–293.

79. Neubauer, C., A. M. Phelan, H. Kues, and D. G. Lange. 1990. Microwave irradiation of rats at 2.45 GHz activates pinocytotic-like uptake of tracer by capillary endothelial cells of cerebral cortex. *Bioelectromagnetics* 11:261–268.

80. Fritze, K., C. Sommer, B. Schmitz, G. Mies, K. A. Hossmann, M. Kiessling, and C. Wiessner. 1997. Effect of global system for mobile communication (GSM) microwave exposure on blood-brain barrier permeability in rat. *Acta Neuropathol.* 94:465–470.

81. Conrad, R. 2013. Smart Meter Health Effects Survey and Report-Exhibit D.

82. Franke, H., E. B. Ringelstein, and F. Stögbauer. 2005. Electromagnetic fields (GSM 1800) do not alter blood-brain barrier permeability to sucrose in models *in vitro* with high barrier tightness. *Bioelectromagnetics* 26(7):529–535.

83. Lai, H. and N. P. Singh. 1995. Acute low-intensity microwave exposure increases DNA single-strand breaks in rat brain cells. *Bioelectromagnetics* 16(3):207–210.

84. Lai, H. and N. P. Singh. 1996. Single- and double-strand DNA breaks in rat brain cells after acute exposure to radiofrequency electromagnetic radiation. *Int. J. Radiat. Biol.* 69(4):513–521.

85. Lai, H. and N. P. Singh. 1997. Melatonin and a spin-trap compound block radiofrequency electromagnetic radiation-induced DNA strand breaks in rat brain cells. *Bioelectromagnetics* 18(6):446–454.

86. Malyapa, R. S., E. W. Ahern, W. L. Straube, E. G. Moros, W. F. Pickard, and J. L. Roti Roti. 1997. Measurement of DNA damage after exposure to 2450 MHz electromagnetic radiation. *Radiat. Res.* 148(6):608–617.

87. Malyapa, R. S., E. W. Ahern, W. L. Straube, E. G. Moros, W. F. Pickard, and J. L. Roti Roti. 1997. Measurement of DNA damage after exposure to electromagnetic radiation in the cellular phone

communication frequency band, (835.62 and 847.74 MHz). *Radiat. Res.* 148(6):618–627.

88. Aitken, R. J., L. E. Bennetts, D. Sawyer, A. M. Wiklendt, and B. V. King. 2005. Impact of radio frequency electromagnetic radiation on DNA integrity in the male germline. *Int. J. Androl.* 28(3):171–179.

89. Markova, E., L. Hillert, L. Malmgren, B. R. Persson, and I. Y. Belyaev. 2005. Microwaves from GSM mobile telephones affect 53BP1 and gamma-H2AX foci in human lymphocytes from hypersensitive and healthy persons. *Environ. Health Perspect.* 113(9):1172–1177.

90. Caraglia, M., M. Marra, F. Mancinelli, G. D'Ambrosio, R. Massa, A. Giordano, A. Budillon, A. Abbruzzese, and E. Bismuto. 2005. Electromagnetic fields at mobile phone frequency induce apoptosis and inactivation of the multi-chaperone complex in human epidermoid cancer cells. *J. Cell. Physiol.* 204(2):539–548.

91. Hook, G. J., P. Zhang, I. Lagroye, L. Li, R. Higashikubo, E. G. Moros, W. L. Straube, W. F. Pickard, J. D. Baty, and J. L. Roti Roti. 2004. Measurement of DNA damage and apoptosis in Molt-4 cells after *in vitro* exposure to radiofrequency radiation. *Radiat. Res.* 161(2):193–200.

92. Capri, M., E. Scarcella, E. Bianchi et al. 2004. 1800 MHz radiofrequency (mobile phones, different global system for mobile communication modulations) does not affect apoptosis and heat shock protein 70 level in peripheral blood mononuclear cells from young and old donors. *Int. J. Radiat. Biol.* 80(6):389–397.

93. Capri, M., E. Scarcella, C. Fumelli et al. 2004. *In vitro* exposure of human lymphocytes to 900 MHz CW and GSM modulated radiofrequency: Studies of proliferation, apoptosis and mitochondrial membrane potential. *Radiat. Res.* 162(2):211–218.

94. Meltz, M. L. 2003. Radiofrequency exposure and mammalian cell toxicity, genotoxicity, and transformation. *Bioelectromagnetics* (Suppl 6):196–213.

95. Cranfield, C. G., H. G. Wieser, and J. Dobson. 2003. Exposure of magnetic bacteria to simulated mobile phone-type RF radiation has no impact on mortality. *IEEE Trans. Nanobiosc.* 2(3):146–149.

96. Chia, S. E., H. P. Chia, and J. S. Tan. 2000. Prevalence of headache among handheld cellular telephone users in Singapore: A community study. *Environ. Health Perspect.* 108(11):1059–1062.

97. Oftedal, G., J. Wilen, M. Sandstrom, and K. H. Mild. 2000. Symptoms experienced in connection with mobile phone use. *Occup Med (Lond)* 50(4):237–245.

98. Santini, R. S. P., P. Le Ruz, J. Danze, and M. Seigne. 2003. Survey study of people living in the vicinity of cellular phone base stations. *Electromagn Biol. Med.* 22(1):41–49.

99. Wilen, J., M. Sandstrom, and K. Hansson Mild. 2003. Subjective symptoms among mobile phone users—a consequence of absorption of radiofrequency fields? *Bioelectromagnetics* 24(3):152–159.

100. Salama, O. E. and R. M. El Naga Abou. 2004. Cellular phones: Are they detrimental? *J. Egypt Public Health Assoc.* 79(3-4):197–223.

101. Al-Khlaiwi, T. and S. A. Meo. 2004. Association of mobile phone radiation with fatigue, headache, dizziness, tension and sleep disturbance in Saudi population. *Saudi Med. J.* 25(6):732–736.

102. Balikci, K., I. Cem Ozcan, D. Turgut-Balik, and H. H. Balik. 2005. A survey study on some neurological symptoms and sensations experienced by long term users of mobile phones. *Pathol Biol (Paris)* 53(1):30–34.

103. Balik, H. H., D. Turgut-Balik, K. Balikci, and I. C. Ozcan. 2005. Some ocular symptoms and sensations experienced by long term users of mobile phones. *Pathol Biol (Paris)* 53(2):88–91.

104. Szyjkowska, A., A. Bortkiewicz, W. Szymczak, and T. Makowiec-Dabrowska. 2005. Subjective symptoms related to mobile phone use—a pilot study. *Pol. Merkur. Lekarski* 19(112):529–532.

105. Meo, S. A., A. M. Al-Drees, S. Husain, M. M. Khan, and M. B. Imran. 2010. Effects of mobile phone radiation on serum testosterone in Wistar albino rats. *Saudi Med. J.* 30(8):869–873.

106. Soderqvist, F., M. Carlberg, and L. Hardell. 2012. Review of four publications on the Danish cohort study on mobile phone subscribers and risk of brain tumors. *Rev. Environ. Health* 27(1):51–58.

107. Landgrebe, M., U. Frick, S. Hauser, G. Hajak, and B. Langguth. 2009. Association of tinnitus and electromagnetic hypersensitivity: Hints for a shared pathophysiology? *PLOS ONE* 4(3):e5026 (1–6).

108. Hutter, H. P., H. Moshammer, P. Wallner et al. 2010. Tinnitus and mobile phone use. *Occup. Environ. Med.* 67(12):804–808.

109. ICNIRP. 1998. Guidelines for limiting exposure to time-varying electric, magnetic and electromagnetic fields (up to 300 GHz). *Health Phys.* 74:494–522.

110. Fragopoulou, A. F., S. L. Koussoulakos, and L. H. Margaritis. 2010. Cranial and postcranial skeletal variations induced in mouse embryos by mobile phone radiation. *Pathophysiology* 17(3):169–177.

111. Fragopoulou, A. F., P. Miltiadous, A. Stamatakis et al. 2010. Whole body exposure with GSM 900 MHz affects spatial memory in mice. *Pathophysiology* 17(3):179–187.

112. Ferrie, H. 2013. *Creative Outrage*. Caledon, Canada: KOS.
113. Macrae, F. 2010. Mobile Phone Users 'Five Times More Likely to Develop a Brain Tumour'. *Mail Online*. Retrieved 2015-12-2. http://www.dailymail.co.uk/health/article-1286665/Mobile-phone-users-times-likely-develop-brain-tumour.html
114. Khurana, V. G., L. Hardell, J. Everaert, A. Bortkiewicz, M. Carlberg, and M. Ahonen. 2010. Epidemiological evidence for a health risk from mobile phone base stations. *Int. J. Occup. Environ. Health* 263–267.
115. Black, D. R. and L. N. Heynick. 2003. Radiofrequency (RF) effects on blood, cells, cardiac, endocrine and immunological functions. *Bioelectromagnetics* (Suppl 6):S187–S195.
116. Vecchia, P., R. Matthes, G. Ziegelberger, J. Lin, R. Saunders, and A. Swerdlow. 2009. Exposure to high frequency electromagnetic fields, biological effects and health consequences (100 KHz–300 GHz). *International Commission on Non-Ionizing Radiation Protection*. 16.2009.
117. Grigor'ev IuG. 2003. Biological effects of mobile phone electromagnetic field on chick embryo (risk assessment using the mortality rate). *Radiats Biol. Radioecol.* 43(5):541–543.
118. Xenos, T. D. and I. N. Magras. 2003. Low power density RF radiation effects on experimental animal embryos and foetuses. In: Stavroulakis, P. (Ed.). *Biological Effects of Electromagnetic Fields*, Springer, pp. 579–602.
119. Belyaev, I. Y., C. B. Koch, O. Terenius et al. 2006. Exposure of rat brain to 915 MHz GSM microwaves induces changes in gene expression but not double stranded DNA breaks or effects on chromatin conformation. *Bioelectromagnetics* 27(4):295–306.
120. Nittby, H., G. Grafström, D. P. Tian et al. 2008. Cognitive impairment in rats after long-term exposure to GSM-900 mobile phone radiation. *Bioelectromagnetics* 29(3):219–232.
121. Lai, H. 1994. Neurological effects of radiofrequency electromagnetic radiation. In: Lin, J. C. (Ed.). *Advances in Electromagnetic Fields in Living Systems*, New York: Plenum Press, vol. 1, pp. 17–88.
122. Belyaev, I. 2010. Dependence of non-thermal biological effects of microwaves on physical and biological variables: Implications for reproducibility and safety standards. In: Giuliani, L. and Soffritti, M. (Ed.). *European Journal of Oncology—Library, vol. 5 Non-Thermal Effects and Mechanisms of Interaction between Electromagnetic Fields and Living Matter. An ICEMS Monograph*, Bologna, Italy: Ramazzini Institute, pp. 187–218, http://www.icems.eu/papers.htm? f¼/c/a/2009/12/15/MNHJ1B49KH.DTL

123. Volkow, N. D., D. Tomasi, G. J. Wang et al. 2011. Effects of cell phone radiofrequency signal exposure on brain glucose metabolism. *JAMA* 305(8):808–813.
124. Lai, H. and L. Hardell. 2011. Cell phone radiofrequency radiation exposure and brain glucose metabolism. *JAMA (Editorial)* 305(8).
125. Zaheer, A., S. Zaheer, S. K. Sahu et al. 2007. A novel role of glia maturation factor: Induction of granulocyte-macrophage colony-stimulating factor and pro-inflammatory cytokines. *J. Neurochem.* 101(2):364–376.
126. Ammari, M., E. Brillaud, C. Gamez et al. 2008. Effect of a chronic GSM 900 MHz exposure on glia in the rat brain. *Biomed. Pharmacother.* 62(4):273–281.
127. Hardell, L. and M. Carlberg. 2009. Mobile phones, cordless phones and the risk for brain tumours. *Int. J. Oncol.* 35(1):5–17.
128. Khurana, V. G., C. Teo, M. Kundi et al. 2009. Cell phones and brain tumors: A review including the long-term epidemiologic data. *Surg. Neurol.* 72(3):205–214.
129. Morgan, T. E., I. Rozovsky, S. K. Goldsmith et al. 1997. Increased transcription of the astrocyte gene GFAP during middle-age is attenuated by food restriction: Implications for the role of oxidative stress. *Free Radic. Biol. Med.* 23(3):524–528.
130. Meral, I., H. Mert, N. Mert et al. 2007. Effects of 900-MHz electromagnetic field emitted from cellular phone on brain oxidative stress and some vitamin levels of guinea pigs. *Brain Res.* 1169:120–124.
131. Nittby, H., A. Brun, J. Eberhardt et al. 2009. Increased blood-brain barrier permeability in mammalian brain 7 days after exposure to the radiation from a GSM-900 mobile phone. *Pathophysiology* 16(2–3):103–112.
132. Sirav, B. and N. Seyhan. 2009. Blood-brain barrier disruption by continuous-wave radio frequency radiation. *Electromagn. Biol. Med.* 28(2):215–222.
133. Fragopoulou, A. F. and L. H. Margaritis. 2010. Is cognitive function affected by mobile phone radiation exposure? In: Giuliani, L. and Soffritti, M. (Ed.). *European J. Oncology-Library, vol. 5 Non-Thermal Effects and Mechanisms of Interaction between Electromagnetic Fields and Living Matter. An ICEMS Monograph*, Bologna, Italy: Ramazzini Institute, pp. 261–272.
134. Ntzouni, M. P., A. Stamatakis, F. Stylianopoulou, and L. H. Margaritis. 2011. Short term memory in mice is affected by mobile phone radiation. *Pathophysiology* 18(3):193–199.

135. Andrews-Zwilling, Y., N. Bien-Ly, Q. Xu et al. 2010. Apolipoprotein E4 causes age- and Tau-dependent impairment of GABAergic interneurons, leading to learning and memory deficits in mice. *J. Neurosci.* 30(41):13707–13717.

136. Schüz, J., E. Böhler, G. Berg et al. 2006. Cellular phones, cordless phones, and the risks of glioma and meningioma (Interphone Study Group, Germany). *Amer. J. Epidemiol.* 163(6):512–520.

137. Söderqvist, F., M. Carlberg, M. K. Hansson, and L. Hardell. 2009. Exposure to an 890-MHz mobile phone-like signal and serum levels of S100B and transthyretin in volunteers. *Toxicol. Lett.* 189(1):63–66.

138. Divan, H. A., L. Kheifets, C. Obel, and J. Olsen. 2008. Prenatal and postnatal exposure to cell phone use and behavioral problems in children. *Epidemiology* 19(4):523–529.

139. Blackman, C. 2009. Cell phone radiation: Evidence from ELF and RF studies supporting more inclusive risk identification and assessment. *Pathophysiology* 16(2–3):205–216.

140. Nittby, H., B. Widegren, M. Krogh et al. 2008. Exposure to radiation from global system for mobile communications at 1,800 MHz significantly changes gene expression in rat hippocampus and cortex. *Environmentalist* 28(4):458–465.

141. Agarwal, A., N. R. Desai, K. Makker et al. 2009. Effects of radiofrequency electromagnetic waves (RF-EMW) from cellular phones on human ejaculated semen: An *in vitro* pilot study. *Fertil. Steril.* 92(4):1318–1325.

142. De Iuliis, G. N., R. J. Newey, B. V. King, and R. J. Aitken. 2009. Mobile phone radiation induces reactive oxygen species production and DNA damage in human spermatozoa *in vitro. PLOS ONE* 4(7):e6446.

143. Irmak, M. K., E. Fadillioglu, M. Gulec et al. 2002. Effects of electromagnetic radiation from a cellular telephone on the oxidant and antioxidant levels in rabbits. *Cell Biochem. Funct.* 20(4):279–e6283.

144. Friedman, J., S. Kraus, Y. Hauptman et al. 2007. Mechanism of short-term ERK activation by electromagnetic fields at mobile phone frequencies. *Biochem. J.* 405(3):559–568.

145. Lee, K. S., J. S. Choi, S. Y. Hong et al. 2008. Mobile phone electromagnetic radiation activates MAPK signaling and regulates viability in *Drosophila. Bioelectromagnetics* 29(5):371–379.

146. Minelli, T. A., M. Balduzzo, F. F. Milone, and V. Nofrate. 2007. Modeling cell dynamics under mobile phone radiation. *Nonlin. Dyn. Psychol. Life Sci.* 11(2):197–218.

147. Blank, M. and R. Goodman. 2009. Electromagnetic fields stress living cells. *Pathophysiology* 16(2–3):71–78.

148. Challis, L. J. 2005. Mechanisms for interaction between RF fields and biological tissue. *Bioelectromagnetics* 26(Suppl 7):98–106.
149. McNamee, J. P. and V. Chauhan. 2009. Radiofrequency radiation and gene/protein expression: A review. *Radiat. Res.* 172(3):265–287.
150. Ong, S.-E. and A. Pandey. 2001. Review: An evaluation of the use of two-dimensional gel electrophoresis in proteomics. *Biomolec. Eng.* 18:195–205.
151. Rogers, M. and J. Graham. 2007. Robust and accurate registration of 2-D electrophoresis gels using point-matching. *IEEE Trans. Image Proc.* 16(3):624–635.
152. Földi, I., G. Müller, B. Penke, and T. Janáky. 2011. Characterisation of the variation of mouse brain proteome by two-dimensional electrophoresis. *J. Proteomics* 74(6):894–901.
153. Kar, P., C. Nelson, and A. B. Parekh. 2011. Selective activation of the transcription factor NFAT1 by calcium microdomains near Ca2 þ release-activated Ca2 þ (CRAC) channels. *J. Biol. Chem.* 286(17):14795–14803.
154. Dart, P., K. Cordes, A. Elliott, J. Knackstedt, J. Morgan, and P. Wible. 2013. Biological and health effects of microwave radio frequency transmissions. *Rev. Res Lit.*
155. Kim, A., J. Chonda, and Law Department PGE. Pacific Gas and Electric Company's Response to Administrative Law Judge's October 18, 2011 Ruling Directing It to File Clarifying Radio Frequency Information. 2011/11/1; http://sunroomdesk.com/wpcontent/uploads/2011/11/PGERResponsesRFDataOpt-outalternatives_11-1-11-3pm.pdf
156. Wilner, D. and Wilner & Associates vs. Pacific Gas and Electric Company. Before the California Public Utilities Commission of the State of California. 2011/10/26; 1–19. http://docs.cpuc.ca.gov/published/proceedings/C1110028.htm
157. Levitt, B. and H. Lai. 2010. Biological effects from exposure to electromagnetic radiation emitted by cell tower base stations and other antenna arrays. *Environ. Rev.* 18:369–395.
158. Yakymenko, I., E. Sidorik, S. Kyrylenko, and V. Chekhun. 2011. Long-term exposure to microwave radiation provokes cancer growth: Evidences from radars and mobile communication systems. *Exp. Oncol.* 33(2):62–70.
159. Altpeter, E. S., M. Roosli, M. Battaglia, D. Pfluger, C. E. Minder, and T. Abelin. 2006. Effect of short-wave (6–22 MHz) magnetic fields on sleep quality and melatonin cycle in humans: The Schwarzenburg shut-down study. *Bioelectromagnetics* 27(2):142–150.

160. BioInitiative Report: A Rationale for a Biologically-Based Public Exposure Standard for Electromagnetic Fields (ELF and RF). *The BioInitiative Report RSS*. Retrieved 2015-12-3. http://www.bioinitiative.org/

161. Repacholi, M., Y. Grigoriev, J. Buschmann, and C. Pioli. 2012. Scientific basis for the Soviet and Russian radiofrequency standards for the general public. *Bioelectromagnetics* 33(8):623–633.

162. Hankin, N. 2002. Biological and Health Effects of Microwave Radio Frequency Transmissions, Center for Science and Risk Assessment, Radiation Protection Division, United States Environmental Protection Agency, to Ms. Jane Newton, President. *The EMR Network* 1–3.

163. Irvine, D. S. 1997. Declining sperm quality: A review of facts and hypotheses. *Baillieres Clin Obstet Gynaecol* 11(4):655–671.

164. Johansson, O. 2011. Letter to California Public Utilities Commission (CPUC) re Smart Meters, pp. 1–3. http://www.scribd.com/doc/59738917/Dr-Johansson-s-letter-re-SmartGridSmart-Meter-dangers-to-CPUC-7-9-2011

165. Aringer, L., J. Cunningham, F. Gobba et al. 1997. Possible health implications of subjective symptoms and electromagnetic fields. In: Bergqvist, U. and Vogel, E. (Eds.). *European Commission DG-V 1997:18*, pp. 1–125. https://gupea.ub.gu.se/dspace/bitstream/2077/4156/1/ah1997_19.pdf

166. Irvine, N. 2005. *Definition, Epidemiology and Management of Electrical Sensitivity*. Report for the Health Protection Agency Centre for Radiation, Chemical and Environmental Hazards, Radiation Protection Division, Chilton, Didcot, Oxfordshire, United Kingdom; HPA-RPD-010: 1–42. http://www.hpa.org.uk/web/HPAwebFile/HPAweb_C/1194947416613

167. Snoy, T. 2011. Visant à faire reconnaître les patients atteints d'électro-hypersensibilité. Chambre des Representats de Belgique. Pp. 1–8. http://www.next-up.org/pdf/Proposition_de_Resolution_Therese_Snoy_Visant_a_faire_reconnaitre_les_patients_atteints_d_electro_hypersensibilite_Chambre_des_Repesentants_Belgique_20_07_2011.pdf

168. Parliamentary Assembly, Council of Europe. 2011. Resolution 1815: The potential dangers of electromagnetic fields and their effect on the environment. Pp. 1–4. http://www.cellphonetaskforce.org/wp-content/uploads/2012/01/eres1815.pdf

169. Firstenberg, A. 2001. Radio Wave Packet. Cellular Phone Task Force: 1–8. http://www.goodhealthinfo.net/radiation/radio_wave_packet.pdf

170. Flynn, J. 2015. Stop the Lies! Stop the Corruption! Stop the Fifty Years of Cover-up! Radiation Belongs to Weapons of War—Not in Consumer Products and Smart Meters.

171. Johansson, O. 2009. Disturbance of the immune system by electromagnetic fields—A potentially underlying cause for cellular damage and tissue repair reduction which could lead to disease and impairment. *Pathophysiology*. 16(2–3):157–177.

172. Siegal, D., J. Li, S. J. Hensley, R. Wilson, R. B. Lansofrd, E. Tai and E. VanDyke. 2018. Incidence rates and trends of pediatric cancer in United States 2001–2014. *Poster presentation of the American Society of Pediatric Hematology/oncology Conference*, Pittsburgh, PA. May 25, 2018.

Index

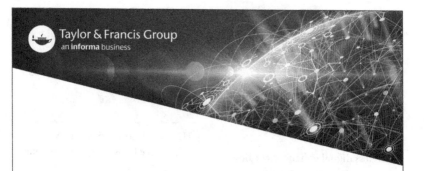

Printed in the United States
by Baker & Taylor Publisher Services